光明城

LUMINOCITY

看见我们的未来

“同济大学学术专著（自然科学类）出版基金”资助项目

王骏阳建筑学论文集

理论·历史·批评(一)

王骏阳　著

同济大学出版社
TONGJI UNIVERSITY PRESS

目录

1

理论何为？

——关于建筑理论教学的思考[1]

1 最初发表于《建筑师》2014 年第 1 期，收录本文集时有修改。

一

“我是教师，算不上理论家”，[1] 这是汪坦先生在20世纪80年中为他主编的“建筑理论译丛”所写前言的开场白。自那时以来，近30年的时间过去了，然而汪坦先生的这句开场白似乎仍然适用，至少对于有着编纂《建筑学论文读本》计划的笔者而言。确实，编纂一部以20世纪下半叶以来的建筑理论为主题的读本，最初的动因就是完成作为教师的教学任务，而且是与所谓“西方建筑理论”教学相关的任务。自从笔者主讲的研究生课程“近现代建筑理论与历史”被确定为“同济大学精品课程”建设项目之后，一部某种意义上的“教材”似乎就在所难免。“精品课程”项目的建设时间是两年，在这两年当中，笔者能够为“教材”做的工作大概也只有一部《建筑学论文读本》，而且还只能是它的第一册，计划中的另外两册要在接下来的两到三年中陆续完成。[2]

上：汪坦主编“建筑理论译丛”
塔夫里《建筑学的理论和历史》（1991年版）
下：汪坦主编“建筑理论译丛”
柯林斯《现代建筑设计思想的演变》（2003年版）

然而，如果汪坦先生的开场白在今天仍然适用，那么与汪坦先生提及的信息闭塞且百废待兴的20世纪80年代相比，今天的中国建筑学界则与整个中国社会同处在一个信息爆炸的时代。除了互联网的信息，建筑理论译著之多几乎可以用海量来形容，其中既有汪坦先生率先开始、之后接踵而至的各种“建筑理论译丛”，也包括种类繁多的其他译著和关于国外理论的介绍文章。另一方面，与20世纪80年代的青年学子对建筑理论的“迫不及待”（汪坦先生语）不同，今天的学生对建筑理论的态度却发生了很大变化——有时完全可以用茫然和拒斥来形容，甚至出现对建筑理论“用脚投票”的状况。[3] 当然，从笔者自己的教学经历来看，今天的学生对建筑理论仍然抱有某种兴趣，不过这种兴趣更多倾向于老师的课程讲授，而不是文本阅读。这或许由

1 汪坦：《前言》，载“建筑理论译丛”，参见曼弗雷多·塔夫里：《建筑学的理论和历史》（郑时龄译），北京：中国建筑工业出版社，1991版，第5页。

2 遗憾的是，这个计划没有能够通过“同济大学精品课程”建设的中期考核，整个《读本》的编纂也随之停顿。按照笔者的理解，中期考核未能通过的原因在于“精品课程”建设的关键内容是教材的编写，而一部以他人论文为主要内容的《读本》在现有学术管理体系内无法被认定为“教材”。

3 青锋、范路：《回归实践——在建筑理论教学上的一些探索》，载《建筑师茶座》/《建筑创作》，2012年第8-9期，第4-7页。

于理论文本本身的晦涩，或许由于外语能力上的不足以及背景知识的缺失。然而，仅仅这些原因似乎并不足以解释今天学生对待建筑理论的态度，因为同样的问题也曾经是80年代“理论热”时期的学生所面临的。除了网络时代所带来的快速信息消费、“微阅读”的习惯和价值取向，或者全民娱乐时代所导致的图像泛滥等很快可以找到的原因，需要进一步思考的应该是这些态度的变化透露了哪些更深层次的问题和挑战？作为建筑学教育的重要组成部分，建筑理论教学应该如何应对这些问题和挑战？

二

理论何为？也许，我们需要一个比“当代理论”更为广阔的历史视野才能更好地回应这个问题。在“西方建筑理论”的语境中，这个“历史视野”可以追溯到古希腊——在那里，“理论”（theoria）一词与“剧院”（theatron）同根同源，从而为“理论”赋予了“看”“观察”“审视”“反思”的基本含义。[1]或者，我们也可以将视野追溯到维特鲁威的《建筑十书》——在那里，“理论”就是知识的系统总结。但是，如果说现代意义上的西方建筑理论始于意大利文艺复兴，那么克里斯多夫·索尼斯（Christof Thones）在他为《建筑理论——从文艺复兴到现在》（*Architectural Theory: From the Renaissance to the Present*）所写的序言中提出的一系列问题无疑值得我们注意，索尼斯开宗明义地发问：“建筑师为什么会写作？他们都写了些什么？何为建筑论著？何为建筑理论？”（What do architects write? And what do they write about? What is a treatise of architecture and what is architectural theory?）尽管索尼斯并没有完全回答这些问题（因为正如索尼斯在提问后随即指出的，对这些问题的回答将充满一个图书馆），但他还是以意大利文艺复兴时期的建筑理论发展为主要线索，涉及这些问题的诸多方面。首先，索尼斯指出，意大利文艺复兴时期的人文主义者致力于将建筑师从“匠人”转变为“文人”，进而参与学者、诗人和知识分子的讨论，而理论则是这一转变的重要途径。其次，与绘画、文学等仅凭艺术家自己就可以“闭门造车”的艺术形式不同，建筑学是一门更为社会化的艺术，需要业主、投资者、权贵、设计者、建造者、使用者的参与和沟通，而“理论”或者更准确地说建筑论著就是为这种沟通提供的工具。事实上，索尼斯告诉我们，作为意大利文艺复兴建筑理论之父的阿尔伯蒂的《建筑十书》（*De re aedificatoria libri decem*）就是献给教皇尼古拉斯五世的。按照索尼斯的观点，文艺复

1 雷蒙·威廉斯：《关键词——文化与社会的词汇》（刘建基译），北京：生活·读书·新知三联书店，2005年版，第486页。

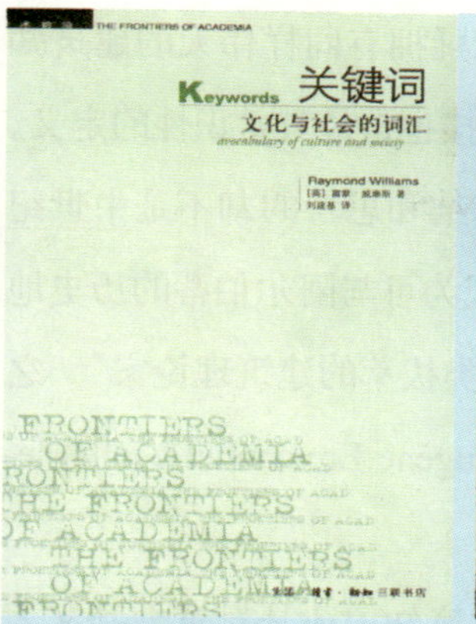

雷蒙·威廉斯《关键词——文化与社会的词汇》将“理论”（theoria）一词的词源追溯到古希腊的“剧场”（theatron）

左：维特鲁威《建筑十书》中文版　中：索尼斯《建筑理论——从文艺复兴到现在》英文版
右：《建筑论：阿尔伯蒂建筑十书》英文版

兴“建筑理论”的第三个基本动因是打破中世纪师徒相授的知识传播形式，使建筑学成为一门通过理论知识进行传授的学科。[1] 一定程度上，这不仅是维特鲁威《建筑十书》的要义，而且也是时至今日建筑理论仍然是建筑学教育重要组成部分的原因之一。

在汉诺-沃尔特·克鲁夫特（Hanno-Walter Kruft）的《建筑理论史——从维特鲁威到现在》（*A History of Architectural Theory: From Vitruvius to the Present*）中，我们不难发现中世纪的建筑理论十分有限。[2] 这一现象也许不奇怪，因为中世纪没有现代意义上的建筑师职业，也没有与之相关的建筑学教育。就此而言，建筑理论与建筑师职业和建筑学教育是相辅相成的。当然，如果说中世纪的欧洲缺少自己的建筑理论，那么这并不意味着那个时代没有自己的建筑知识。恰恰

沃尔特·克鲁夫特
《建筑理论史——从维特鲁威到现在》中文版

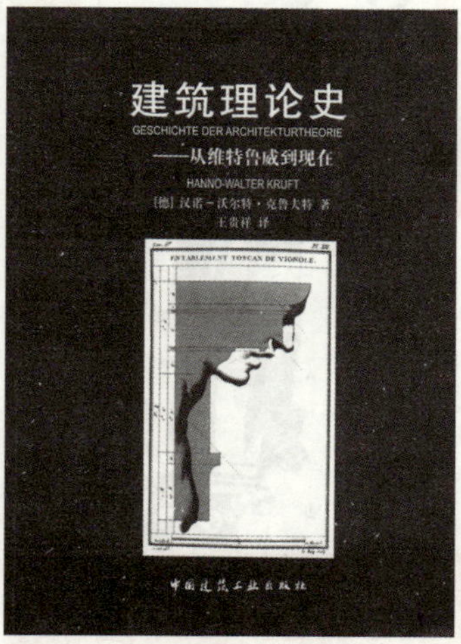

1 Christof Thoenes, “Introduction,” *Architectural Theory: From the Renaissance to the Present* (Cologne: Taschen, 2003), pp.8-19 (8).

2 汉诺-沃尔特·克鲁夫特：《建筑理论史——从维特鲁威到现在》（王贵祥译），北京：中国建筑工业出版社，2005 年版，第二章“维特鲁威的传统与中世纪的建筑理论”。

相反，创造了伟大的哥特教堂的欧洲中世纪无疑拥有同样伟大的建筑知识。理论是知识的系统总结，这是“理论”最为直接也是最具常识性的定义。然而有趣的是，真正对中世纪建筑知识进行总结和思考的却不是中世纪本身，而是 19 世纪，特别是曾被建筑史学家视为可与阿尔伯蒂的历史地位并驾齐驱、被称作“欧洲建筑史上两位最出类拔萃的建筑理论家”[1]之一的尤金 · 伊曼纽尔 · 维奥莱 - 勒 - 迪克（Eugène Emmanuel Viollet-le-Duc，1814—1879）的建筑论著。

但是，尽管缺少系统的总结，中世纪的建筑知识还是对意大利文艺复兴建筑产生了巨大影响。也许没有什么能够比伯鲁乃列斯基设计的、在我国外建史教科书上有文艺复兴“第一朵报春花”之誉[2]的佛罗伦萨主教堂更能说明这一点了。事实上，如果没有中世纪的建筑知识以及伯鲁乃列斯基对这些知识的继承和发展，这个教堂的建造是不可想象的。知识 —— 我这里特别指的是建造知识 —— 不是作为系统性理论，而是作为实践性的技艺以经验的方式传播、继承和发展，这既是欧洲中世纪的状况，也是许多传统人类社会和文化的共同特征。同样，在中国数千年的建筑发展中，与受到“法式”规定的官式建筑不同，大多数民间建筑基本上都遵循着以经验为主的操作路线，尽管不可否认的是，自从姚承祖的《营造法原》问世之后，《营造法式》与地方建筑传统的关系一直被许多学者所关注。[3] 20 世纪初，随着中国现代建筑学的诞生，以“营造学

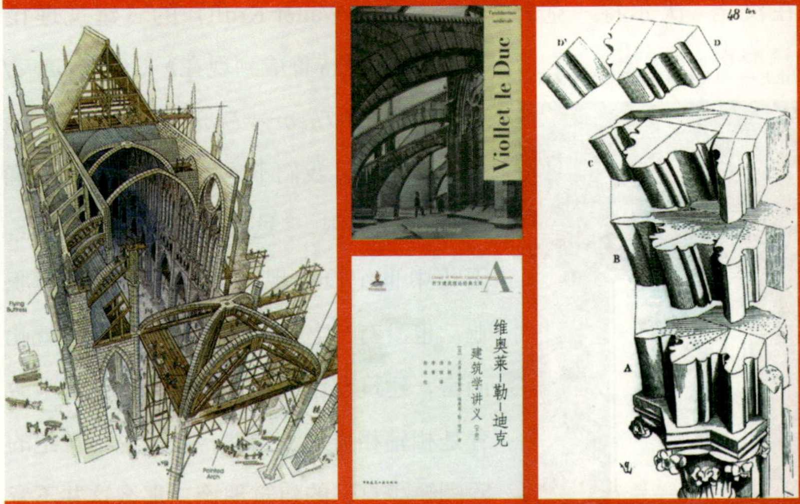

左：建造中的哥特教堂　中上：维奥莱 - 勒 - 迪克《中世纪建筑词典》法文版　中下：《建筑学讲义》中文版
右：维奥莱 - 勒 - 迪克著作中的 12 世纪建造示例

1 John Summerson, “Viollet-le-Duc and the Rationalist Point of View,” *Heavenly Mansions and other Essays on Architecture* (New York and London: W. W. Norton & Company, 1963), pp.135-158 (135).

2 陈志华：《外国建筑史（19 世纪末叶以前）》，北京：中国建筑工业出版社，2004 年版，第 134 页。

3 项隆元：《〈营造法式〉与江南建筑》，杭州：浙江大学出版社，2009 年版，第 28 页。

伯鲁乃列斯基
佛罗伦萨主教堂

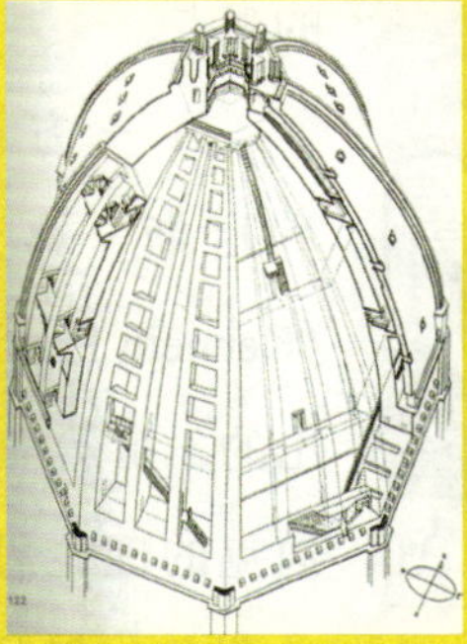

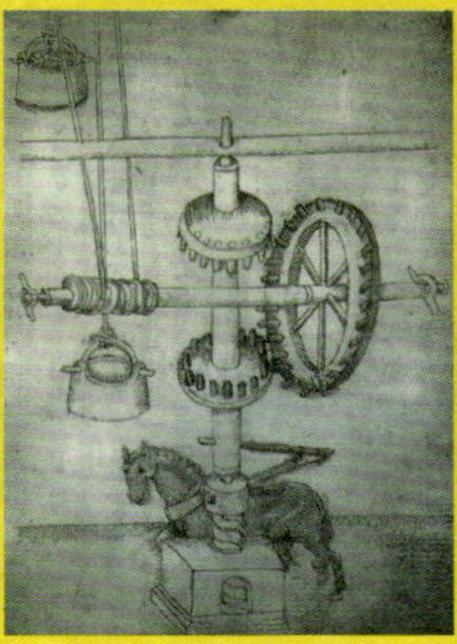

左：佛罗伦萨主教堂穹顶解剖图　中：穹顶施工中使用的以一匹马来驱动的“牛吊车”
右：罗斯·金《圆顶的故事》中文繁体版

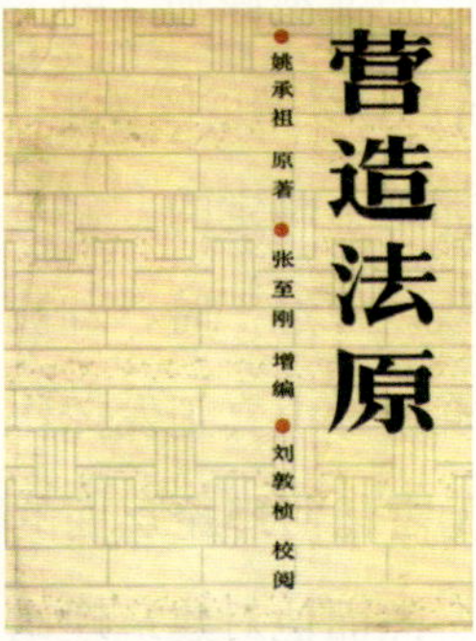

上：姚承祖《营造法原》
下：《中国营造学社会刊》

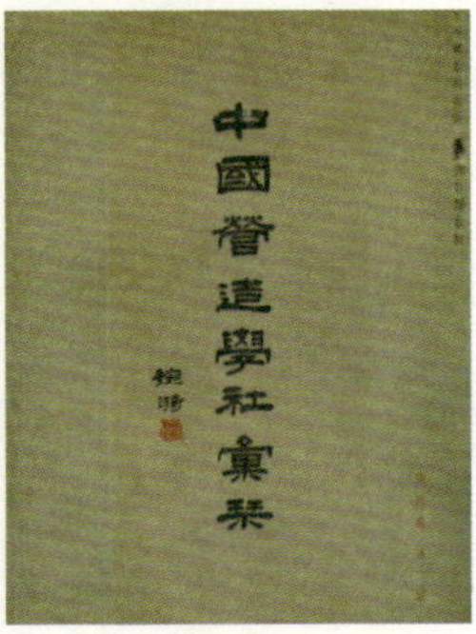

社”为代表的建筑理论研究蓬勃兴起，形成了中国现代建筑学理论研究的第一个高潮。在西方建筑中，正如我们已经看到的，这一变化始于意大利文艺复兴。

文艺复兴在建立西方建筑学理论体系并改变建筑知识传播方式的同时，也从一开始就彰显了建筑学作为一种学科在理论和实践两方面面临的悖论。毫无疑问，建筑是一个需要实践知识的行业，但是建筑学又不能局限于实践知识。与此同时，如果说理论是实践知识的系统总结，那么维奥莱-勒-迪克在19世纪对哥特建筑的理论总结已经在时间上大大“滞后”了。然而，建筑学发展史上也有相当多的案例可以说明，理论一旦形成，也会包含超越实践的内容，无论这种超越是在历时还是在共时的意义上而言的。如果还是以维奥莱-勒-迪克的建筑理论为例，那么很显然，这一理论并非仅仅是对哥特建筑的总结，而是对哥特建筑的诠释，并且正是以这样的诠释为基础，维奥莱-勒-迪

维奥莱－勒－迪克音乐演奏大厅设想

克提出了他的“三千人音乐大厅”的理论设想以及被后人总结为“结构理性主义”（Structural Rationalism）的建筑原则。就此而言，我们应该同意理论是“对实践提出解释的一种思想体系”（a scheme of ideas which explains practice）[1]的观点。

另一方面，正如肯尼斯·弗兰姆普敦（Kenneth Frampton）指出的，虽然维奥莱-勒-迪克的理论思想产生于19世纪，但是它的真正影响却在20世纪，而且无论在19世纪还是20世纪，能够名副其实成为维奥莱-勒-迪克思想传人的建筑师和理论家屈指可数。[2]关于理论与实践之间这种时而滞后时而超前、时而相辅相成、时而又彼此独立的关系，英国建筑史学家约翰·萨默森（John Summerson）曾经精辟阐述道：“建筑理论与建筑生产的实际关系在任何一个时代都是不确定的。完全可能出现这样的情况：一种新思想在某一个时代出现，风靡一时，却被下一个时代肆意践踏，而只有到第三个时代才在建筑设计中真正发挥作用。…… 另一方面，建筑风格的变革也完全可能无需任何理论层面的姿态。还有谁比米开朗基罗对建筑产生的影响更大？然而他对建筑理论的贡献几乎为零。因此，我们必须牢记，理论是一种历史过程，它有自己的生命，存在于自己的文字之中，而且根本就没有什么完全从实践中概括而来的原理，也没有必然反映理论的实践。”[3]

文艺复兴致力于将建筑学提升为人文学科，这也为建筑学转变成为一种思想和观念的学科（而非仅仅是建造技术或实用学科）打开了大门。学者们注意到，在意大利文艺复兴时期丰富繁杂的理论论著中，除

1 雷蒙·威廉斯：《关键词——文化与社会的词汇》，第487页。

2 肯尼斯·弗兰姆普敦：《建构文化研究——论19世纪和20世纪建筑中的建造诗学》（王骏阳译），北京：中国建筑工业出版社，2007年版，第52-62页。

3 John Summerson," The Case for a Theory of Modern Architecture," *Architecture Culture 1943-1968*. A Documentary Anthology, ed. Joan Ockman (New York: Rizzoli, 1993), pp. 227-236 (228).

了阿尔伯蒂的《建筑十书》这类“学术正统”之外，还有诸如《寻爱绮梦》(*Hypnerotomachia Poliphili*)，或者根据后来的英文译本的标题又被称为《波里菲利之梦——梦中爱的纷争》(*Hypnerotomachia Poliphili — The Strife of Love in a Dream*) 的“另类”建筑学著作。这部著作记述的是主人公波里菲利 (Poliphili，也可解作“许多事情的爱人”的意思) 在寻求他的爱人波莉娅 (Polia，也可解作“许多事情”的意思) 的探险过程中经历的如梦似幻的销魂世界和性爱体验。关于这部建筑学奇书的作者，史学家们众说纷纭，莫衷一是，有说无名氏的，也有说是来自威尼斯的方济各会 (Ordine Francescano) 修士弗朗西斯科·科罗纳 (Francesco Colonna) 的。20 世纪 90 年代，更有学者经过研究后语出惊人：作为对《建筑十书》的补充，该书的作者不是别人，正是奠定现代西方建筑学理论基础的阿尔伯蒂！[1] 在此，我们没有可能对这一观点的可靠性进行论证，但是倘若史实确实如此，我们便有理由认为，在阿尔伯蒂那里，或者更准确地说应该是在该书的作者那里，建筑理论或者说理论行为绝非仅仅是对已有知识的系统总结或者对建筑原理的思考，也不是为了建造者、参与者之间的交流，或者使建筑学成为通过理论进行传授的学科，而是一种为满足作者的智识需要而进行的思辨活动。只是，与《建筑十书》的“理性”思维不同，《波里菲利之梦》更强调身体和感性对于思想的重要。20 世纪现象学对笛卡尔理性主义展开批判，或者当代文化理论将想象和欲望置于与理性同等重要的地位，其根源在《波里菲利之梦》中似已可见一斑。在建筑界，曾经以一部《建筑与现代科学的危机》(*Architecture and the Crisis of Modern Science*) 而著称的墨西哥裔加拿大学者阿尔贝托·佩雷兹-戈麦兹 (Alberto Pérez-Gómez) 以《波里菲利之梦》为切入

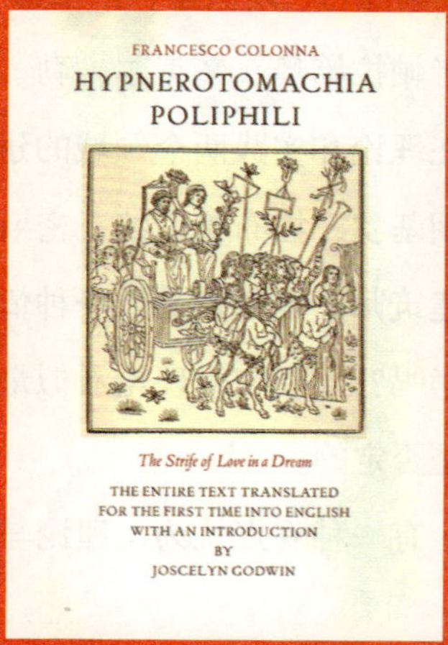

弗朗西斯科·科罗纳《波里菲利之梦——梦中爱的纷争》英文版

1 见 Liane Lefaivre, *Leon Battista Alberti's Hypnerotomachia Poliphili: Re-cognizing the Architectural Body in the Early Italian Renaissance* (Cambridge, Massachusetts and London：The MIT Press, 1997).

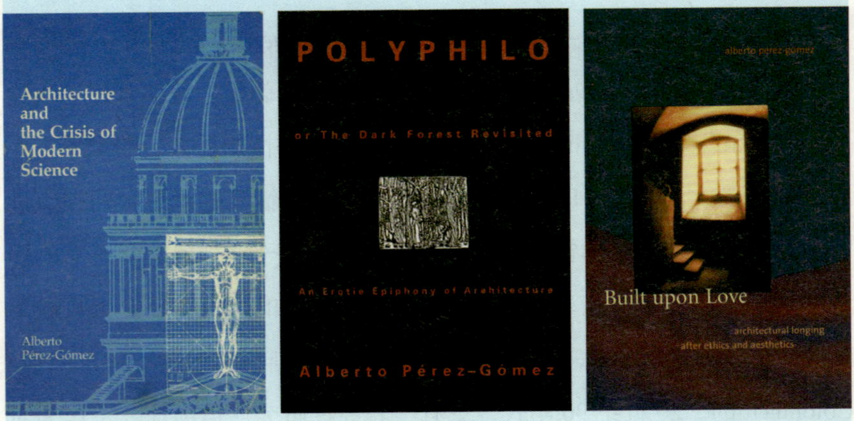

阿尔贝托 · 佩雷斯 - 戈麦兹
左：《建筑学与现代科学的危机》英文版　中：《波里菲利——重访黑森林》英文版　右：《基于爱的建造》英文版

点，提出建筑现象学的"爱欲空间"（erotic space）概念，进而倡导"基于爱的建造"（Built upon Love）的建筑伦理和美学也就没有什么奇怪。[1]

然而，对于本文而言，《波里菲利之梦》的重要之处并不在于它的现象学意义或文化理论意义（尽管这些意义的重要性不言而喻），而是它比《建筑十书》更显著地体现了文艺复兴之后的西方建筑理论的两个发展倾向：首先是建筑理论中越来越多的思辨性，其次是理论与实践的相对独立性。在这里，思考的主体是个体，而非集体（或者说集体是由许许多多独立思考的个体组成的）或者官方，而且这种思考本质上是争辩性的或批判性的。这些思考可以由建筑师通过具体的建筑设计来进行，也可以由建筑师或理论家以文字或绘图来完成。因此，在意大利文艺复兴以来的西方建筑学中，我们不仅看到阿尔伯蒂、洛吉耶、维奥莱 - 勒 - 迪克、勃隆台、塔夫里、诺伯格 - 舒尔茨、柯林·罗这样的"专业"理论家，或者帕拉第奥、辛克尔、勒 · 柯布西耶、赖特、文丘里、库哈斯这样活跃在理论和实践两个领域的建筑师，还有伯鲁乃列斯基、伯拉孟特、米开朗基罗、拉布鲁斯特、密斯、赫佐格与德默隆、盖里等"拒绝"理论的建筑师。但是在最后一种情况下，我们能够因此认为这些建筑师拒绝思考吗？即便在对理论看似最"不屑一顾"的盖里那里，答案或许也应该是否定的。

有一种观点认为，理论与实践的关系应该是对应的，而且对应的程

1 见 Alberto Pérez-Gómez, *Polyphilo: or The Dark Forest Revisited - An Erotic Epiphany of Architecture* (Cambridge, Massachusetts and London：The MIT Press, 1992), and *Built upon Love: Architectural Longing after Ethics and Aesthetics*, (Cambridge, Massachusetts and London：The MIT Press, 2006).

度越高，越能说明理论的有效性。[1]然而，在笔者看来，即使单就同时活跃在理论和实践两个领域的建筑师的情况而言，与其不遗余力寻找他们的理论和实践之间的对应关系，不如将这些建筑师的理论和实践视为思考建筑的不同形式更为恰当。二者之间可能有呼应或对等关系，但并非必然如此。换言之，与理论是知识的系统总结，或者理论是一种对事物的解释或诠释的观念不同，理论还可以是一种反思建筑的方式，而且正是在这样的反思过程中，理论获得了它与实践的相对独立性和自主性甚至对立性。

值得注意的是，尽管维奥莱 - 勒 - 迪克的建筑理论超越了对知识的系统总结，成为一种反思甚至是批判性反思，但是他的理论思考仍然努力围绕建筑的物质本体展开，就像他那个时代的“组合”（composition）理论和“类型”理论也是建立在建筑的物质本体之上的一样。用今天的话来说，它是对建筑学的“自主”研究。进入 20 世纪以来，这种“学科自主”逐步被形式语言、风格派、九宫格教学法以及艾森曼建立在“自我指涉”的符号和形式的“内在性”（interiority）基础之上的“批判建筑学”所取代，而以弗兰姆普敦的《建构文化研究》（*Studies in Tectonic Culture*）为代表的理论倾向则试图在使建筑学重回维奥莱 - 勒 - 迪克式的建筑本体的同时，努力维系建筑学的诗性维度和社会批判的可能。相比之下，《波里菲利之梦》则致力于通过脱离这种“本体”来拓展建筑学领域的空间和边界。这种趋势和努力在文艺复兴之后的西方建筑学发展之中一直若隐若现，并在 20 世纪演变为对空间体验的诉求和对建筑的现象学认知。另一方面，对于 20 世纪建筑学特别是当代建筑学而言，“城市”已经在越来越大的程度上取代“身体”成为拓展建筑学学科边界的有力资源。在后一个方面，荷兰建筑师雷

上：彼得·艾森曼《艾森曼内转外》英文版
下：《对立面读本》英文版

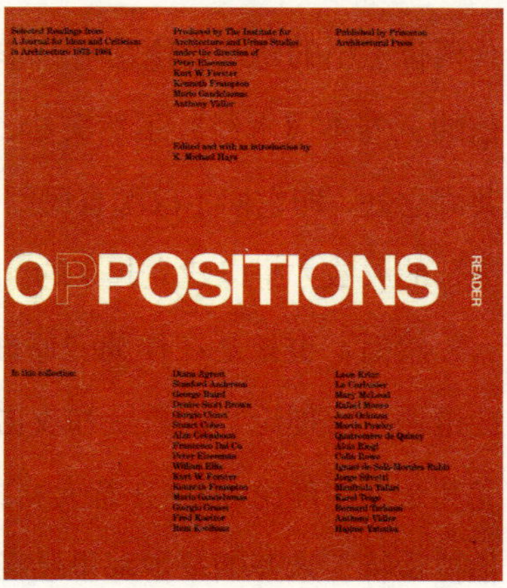

1 青锋、范路：《回归实践——在建筑理论教学上的一些探索》。文中提出，“对于任何理论，最有效的评判方式无疑就是看它所支持的建筑是否达到了理论所预言的效果”。因此，该文主张将建筑理论教学的重点放在“有代表性建筑作品的理论，如灰派、白派、新理性主义、批判地域主义、建构、以及极端现实主义，而那些没有多少‘建筑’的理论如解构、块茎、褶皱、图解、批判与后批判的争论等内容，以及没有多少‘理论’的建筑如高技派、新现代，以及某些数字化设计等等都仅仅给予相对简要的介绍”。

左：弗兰姆普敦《建构文化研究》中文版　右：库哈斯《小、中、大、特大》英文版

姆·库哈斯（Rem Koolhaas）的理论工作当属出类拔萃之列。

理论常常需要面对原理或原则的问题。在这方面，如果阿尔伯蒂的《建筑十书》或者维奥莱-勒-迪克的《建筑谈话录》（*Entretiens sur l'architecture*）的标题还不够一目了然的话，那么20世纪挪威建筑理论家和史学家克里斯蒂安·诺伯格-舒尔茨（Christian Norberg-Schulz）将自己的最后一部理论著作取名为《现代建筑原则》（*Principles of Modern Architecture*）[1]则很能说明问题。即使在中国古代为数不多的建筑文献中，"法式"的主导地位也在很大程度上发挥着"原则"甚至"社会法则"的作用。有趣的是，在中国传统文化中，尽管画论和史论十分发达，众说纷纭，莫衷一是，然而建筑论著不仅在数量上十分有限，而且大多是以技术规范和满足社会管理需要为特点的"官书"，其中《木经》和《营造法式》的作者喻皓和李诫被张钦楠先生在《中国古代建筑师》一书中称为"宋代两位杰出的建筑理论家"。[2]与李诫不同，喻皓是一个非官方的民间工匠，他的三卷本《木经》虽然失传，但是与明代计成的《园冶》却有着某些共同之处，即它们不仅源自民间，而且其历史和专业地位也绝非依靠"官书"的权威才得以建立起来的。

然而，也正是在作为"原理""原则""法典"的问题上，理论常常会遭遇"反理论"的质疑和攻击。这种"反理论"要么是拒绝一切理论，尤其是文字理论——在意大利文艺复兴时期，这种倾向在伯鲁乃列斯基、伯拉孟特、米开朗基罗那里已经表现得十分明显；[3]要么是提出某种"反理论"的立场，尽管这些"反理论"常常"不过是另一种新理论，只不

---

1 见 Christian Norberg-Schulz: *Principles of Modern Architecture* (London: Andreas Papadakis Publisher, 2000).

2 张钦楠：《中国古代建筑师》，北京：生活·读书·新知三联书店，2008年版，第159页。

3 Christof Thoenes, "Introduction," p.19.

过这些新理论自豪地将旧理论的死亡作为自己的出发点而已”[1]。这一点甚至可以适用于维奥莱 - 勒 - 迪克对巴黎美院理论的批判。同样，《波里菲利之梦》也可被视为这种“反理论”的理论。在这方面，虽然《园冶》没有摆出一副“反理论”姿态，但是这部中国古代建筑文献中最具有“自由思想”意义的论著能够与童寯先生的《江南园林志》一样成为当代王澍“人文建筑学”的思想资源并非巧合。在这样的“人文建筑学”中，“园林”“山水画”“民间建造”都是地方或特殊元素，就如“意大利文艺复兴”或者“哥特建筑”是地方和特殊元素一样，而“思考”“批判”“叛逆”“突破体制”才是理论思维的普遍精神。然而，在王澍那里这一切又都不是“无知者无畏”式的盲动，而是建立在广泛的理论阅读基础之上的。同样值得注意的是，王澍曾经这样说道：理论写作和建筑作品是“一种相互平行、相互刺激、相互影响的关系。我的文章……和我的房子是平行的。……文章是独立的，就像理论是相对独立的一样”。[2]

一定意义上，西方建筑学的发展就是在“理论”与“反理论”的互动中得以向前推进的，而在 2000 年前后出现的那场“批判性”（the critical）与“投射性”（the projective）之争——或者说“理论”与“理论之后”（after theory）之争——只不过是这个发展过程中的最新发展而已。它不仅再一次彰显了西方建筑学的发展特点，也在一定程度上说明了“理论”的本质，因为“如果一个理论公之于世而无人反对的话，那它必定不再是一种理论，而将成为一种规律（law）”[3]。就此而言，也许没有什么人能够像 20 世纪法国哲学家福柯的话更好地阐述了“理论”的

左：张钦楠《中国古代建筑师》　中：陈植《园冶注释》（计成原著）　右：童寯《江南园林志》（1963 年版）

1 青锋、范路：《回归实践——在建筑理论教学上的一些探索》，第 4 页。

2 易娜、李东：《界于理论思辨和技术之间的营造——建筑师王澍访谈》，载《深度报道》/《建筑师》2006 年第 4 期（总第 122 期），第 16-25(21) 页。

3 雷蒙 · 威廉斯：《关键词——文化与社会的词汇》，第 487 页。

这一本质在思想层面的意义："当今的哲学（我指的是哲学活动），如果不是关于它本身的批判性思考，又该是什么呢？而且，如果这种哲学仍只在乎证明人们已知的事物有理，而不在乎去弄清楚怎样和在什么限度内有可能用异样的方式进行思考，它又能是什么？"[1]

当然，理论的这种"自我否定"式的发展也会带来困惑。美国学者乔纳森·卡勒（Jonathan Culler）就曾指出："如今的理论最令人失望的一点在于它永无止境。它不是那种你能够掌握的东西，不是一组专门的文章，你只要读懂了，便'明白了理论'。它是一套包罗万象的文集大全，总是在不停地争论着，因为年轻而又不安分的学者总是在批评他们的长辈们的指导思想，刺进新的思想家对理论做出新的贡献。因此理论就成了一种令人惊恐不安的源头，一种不断推陈出新的资源。…… 理论常常又会像一项残忍的判决，逼着你去阅读你不熟悉的领域中的那些十分难懂的文章。在那些领域里，攻克一部著作带给你的不是短暂的喘息，而是更多的、更艰难的阅读"。[2]

卡勒说的是文学理论（literary theory）的状况，不过同样的问题显然也出现在当代建筑理论之中。一方面，建筑学理论文章和论著的数量与日俱增，如何在如此繁杂的范围中进行筛选对于一部以理论、历史和批评为宗旨的建筑学论文读本的理论视野和诉求无疑形成巨大考验和挑战。另一方面，20 世纪 60 年代以后的建筑理论所呈现的过分哲学化的理论话语（这种情况尤以美国东海岸的建筑理论界为甚）使建筑理论读本回归建筑学基本问题成为必要。说到底，建筑理论需要做的是直面自己学科的问题，而不总是试图用其他学科（尤其是哲学）的话语来支撑和证明

左：卡勒《文学理论入门》中英文双语版　　右：英国《建筑设计》杂志 2009 年专辑《理论的崩溃》

---

1 詹姆斯·米勒：《福柯的生死爱欲》，上海：世纪出版集团 / 上海人民出版社，2003 年版，第 34 页。

2 乔纳森·卡勒：《文学理论入门》（李平译），南京：凤凰出版传媒集团 / 译林出版社，2008 年版，第 16 页。

自己的学科合法性。尽管这一立场并不意味着建筑学应该拒绝与包括哲学在内的其他学科的交融，但是努力展现建筑理论与建筑学基本问题的关系、从而为学生理解更多更新的建筑学问题提供一个基本平台不仅应该是建筑理论教学的基本任务，也应该是一部建筑理论读本应该追求的目标。

在最近的“反理论”浪潮中，迈克尔·斯皮克斯（Michael Speaks）的“理论之后”主张最具代表性，它要求“远离建筑理论，在某些情况下甚至拒绝建筑理论”[1]。尽管斯皮克斯自己倡导的“设计智慧”（design intelligence）[2]或者提出“投射性建筑学”一词的罗伯特·索莫（Robert Somol）和萨拉·怀汀（Sara Whiting）所主张的“多普勒效应”（Doppler effect）[3]都完全可以说是一种“理论”，但是斯皮克斯的主张还是将一种“反理论”立场表述得淋漓尽致。他写道：“理论不但无关紧要，而且过去是，还将继续是建筑创新文化发展的一个障碍物。理论先是摆出一副新生姿态，以替代启蒙运动的必然事物，但这其实又使它自身反过来沦为老套的启蒙式批判，这样的批判永不疲倦地一茬跟着一茬，或漫无目的，或没完没了。理论同时也使启蒙运动的这种信念生生不息，即思想是脱离于实际做法，又是指导实际的，也就是说，宣言指导政治行动，建筑理论指导建筑实践”。[4]

斯皮克斯的论点看似掷地有声，然而它在将“反理论”主张推向极致之时，似乎也泄露了这一主张的“天机”：原来，新锐的“反理论”立场只不过是一次老调重弹——“理论指导实践”。或许，这就是斯皮克斯对理论的认识。一定程度上，似乎正是这种“理论指导实践”的立场本身导致了“理论的崩溃”（theoretical meltdown），如果我们借用英国建筑杂志《建筑设计》（*Architectural Design*）在一期反理论专题中使用的标题的话。[5]另一方面，正如关注“批判性”与“投射性”之争的建筑学者乔治·贝尔德（George Baird）指出的：“这场有趣的观点的分化所带来的政治联盟和理论的复杂多变，并未形成这一故事的结束，而仅仅是开始。在一个真正强健的、经得起考验的新的职业立场建立之前，还有许多真正重要的问题需要解答。…… 最根本的是，一个新的‘投射的建筑学’不

---

1 迈克尔·斯皮克斯：《理论之后》（周鸣浩译），载《时代建筑》2007年第5期（总第97期），第72-73(73)页。

2 迈克尔·斯皮克斯：《设计智慧与新经济》（凌琳译），载《时代建筑》2007年第4期（总第96期），第80-83页。

3 罗伯特·索莫、萨拉·怀汀：《关于“多普勒效应”的笔记和现代主义的其它心境》（范凌译），载《时代建筑》2007年第2期（总第94期），第112-117页，以及范凌：《围绕＜关于“多普勒效应”的笔记和现代主义的其它心境＞一文的五点对话》，载《时代建筑》2008年第1期（总第99期），第122-127页。

4 斯皮克斯：《理论之后》，第73页。原文为“theory is not just irrelevant but … ”。见 Michael Speaks, “After Theory: Debate in architectural schools rages about the value of theory and its effect on innovation in design,” *Architectural Record*, 2006-05, pp.73-75(74). 周鸣浩译文为“理论不但不是无关紧要，……”，似有误。

5 见 *Architectural Design*, Vol 79 No 1, Jan/Feb 2009.

伊格尔顿《理论之后》英文版与中文版

能在没有一套‘投射的理论’支持下发展起来。没有这样一套理论，我预言这一新建筑会以惊人的速度沦为‘仅仅’注重实用、‘仅仅’注重装饰的一种建筑”[1]。同样，当代文化学者特里·伊格尔顿（Terry Eagleton）则在他那本同样题为《理论之后》的论著中尖锐指出：“如果这本书的书名表明‘理论’已经终结，我们可以坦然回到前理论的天真时代，本书的读者将感到失望。…… 我们永远不能在‘理论之后’，也就是说没有理论，就没有反省的人生”。[2]

上：拉斯金《建筑七灯》中文版
下：森佩尔《风格论》英文版

伊格尔顿的观点再次表明了“理论”的一个重要性质：与其说它是指导实践的，不如说是反省或者是回应实践的。作为一种反思，理论的首要任务是提出问题，而非提供答案，或者用这种答案指导实践。这一点可以在建筑历史上的伟大理论那里得到诸多印证，因为在某种意义上，这些理论都是在建筑学的历史条件发生重大变化而面临根本挑战之时形成的，它们试图对这些挑战进行反思和回应，提出问题，当然也包括理论者对问题的回答。但是，

1 乔治·贝尔德：《“批判性”及其不满》（都铭译），《时代建筑》2007 年第 3 期（总第 95 期），第 54–57(57) 页。
2 特里·伊格尔顿：《理论之后》（商正译），北京：商务印书馆，2009 年版，第 3 和 213 页。

好的理论的真正价值并不在对问题的回答（说到底，它只是理论者的一家之言），而在于问题本身。19 世纪，当工业革命和新技术形成对建筑学重要挑战之时，约翰 · 拉斯金、维奥莱 - 勒 - 迪克、哥特弗里德 · 森佩尔（Gottfried Semper）的建筑理论无一不是对这些挑战的反思和回应。人们今天也许很难完全认同拉斯金的立场，而维奥莱 - 勒 - 迪克的“结构理性主义”也遭遇建筑表达上的许多困难，更不要说森佩尔的“建筑四要素”常常被简化为一种表皮理论，但是无论作为“答案”的理论在今天会受到怎样的质疑，只要它们试图回应的“问题”仍然存在，那么我们就仍然需要对它们做出思考和回应，而拉斯金、维奥莱 - 勒 - 迪克和森佩尔的理论仍然可以成为一种参照，如果不是“指导”的话。

换言之，如果认为理论的任务是形成原则（理论有时确实是这样的，至少试图这样做），那么“理论指导实践”就是一个不可避免、同时也是一个可疑的结论。在这样的意义上，歌德认为“对于天才来说，原则比案例更危险”[1]，或者斯皮克斯对“理论之后”的向往便不足为奇。然而，如果将理论视为一种反思或者反省，“如果理论意味着对我们的指导性假设进行一番顺理成章的思索，那么它还是一如既往地不可或缺”。[2]

三

对于理论与实践之间的关系，笔者所接触的相当一部分学生的想法也许没有这么复杂，他们只是希望通过理论学习提高自己的设计技巧和设计能力。关于这一点，不妨看看作为建筑师的王澍是怎么说的：“理论可以建造理论的体系，而从来不会说学会了一套理论之后，你就会做建筑了，这是两回事。理论是一种和建筑有关的以文字和脑子里的思想为基础的另外的一个思辨过程，那个过程会对建筑产生说不清道不明的影响，这很重要，但它不能够直接变成方案，永远不可能这样。所以就像很多人做不好建筑，是因为理论太少，不是这样的。完全可以不知道任何理论，就可以做很好的建筑，不一定需要知道的。”[3]

有趣的问题是，既然如此，王澍为什么要进行理论写作？还是那句话，理论是一种不完全等同于设计本身的思考，它提出问题，而不是答案，更不能提供设计技巧和设计能力。如果硬要在理论和设计技巧和设计能力之间建立某种关系，那么应该是理论能够帮助设计者思考如何使用自己的设计技巧和设计能力，前提是你已经具备了这些技巧和能力，但是

1 Christof Thoenes: “Introduction,” p.19.
2 伊格尔顿：《理论之后》，第 3 页。
3 易娜、李东：《界于理论思辨和技术之间的营造——建筑师王澍访谈》，第 21 页。

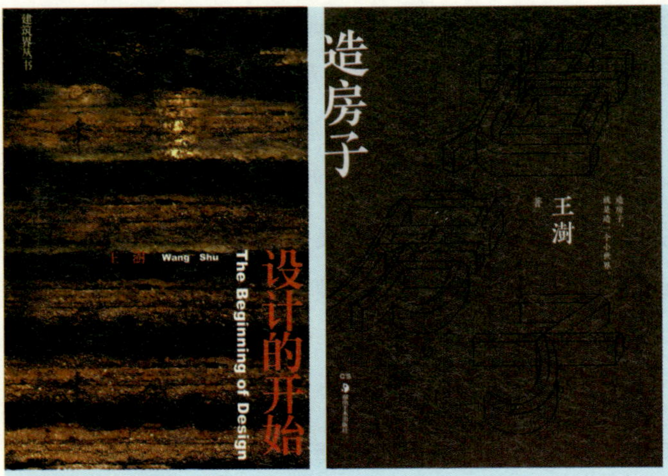

王澍文集《设计的开始》
与《造房子》

这种技巧和能力并非来自理论学习，而是来自从本科学习就已经开始的设计训练和基本功培养。这正是建筑学科的特殊性，也是一些从未真正受过建筑设计训练的理论家们常常会在理论噱头的云山雾罩之中而不知所云的原因之所在吧。

然而，尽管理论无法保证设计技巧和设计能力的获取，理论教学仍然是建筑学教育不可或缺的组成部分。它的基本任务是通过历史视野的构建提高学生对建筑学问题的认识。诚如萨默森所言，当代建筑理论“不是我们时代的武断发明，而是漫长历史演变的一个新阶段”，因此“只有在观念史的层面上才能对建筑理论进行传授”[1]。建筑理论教学的任务还在于培养学生对建筑学问题的批判性反思能力。这种反思需要以建筑学的理论知识为基础，但是理论教学的目的却不能仅仅是传授甚至灌输知识和教条，而要努力激发学生对建筑学问题的思考。笔者相信，无论是作为“观念史”，还是以激发“批判思维”为目的，文本阅读都必不可少，因为只有通过“文本的阅读才能获得相对准确的信息与严谨的认识，以避免一种模糊性的或口号式的学习状况”。[2]

四

最后，有必要对本文伊始提到的“近现代建筑理论与历史”课程的历史和现状做一个简单的介绍。在同济大学建筑与城规学院，这门课是在 20 世纪 80 年代初由罗小未先生创建和开设的。罗先生退休之后，这门课曾转由伍江老师主持，之后又转为卢永毅老师主持。

1 John Summerson: “The Case for a Theory of Modern Architecture”, p.235.

2 卢永毅：《导言》，载《建筑理论的多维视野》（卢永毅主编），北京：中国建筑工业出版社，2009 年版，第 12-20（12）页。

2007 年，笔者到同济大学任教，接手这门课的教学任务。因此，尽管该课的名称是“近现代建筑理论与历史”（这个名称自罗先生开创这门课以来一直沿用至今），但是它却是一门由所谓“外国建筑史”教学团队主持讲授的课程，因而讲课内容多以“外国建筑”尤其是“西方建筑”的问题为主。罗先生写道：“20 世纪 70 年代末 80 年代初，同济大学进入改革开放的年代，建筑教学开始走上正轨，长期处于政治批判漩涡中的外国建筑历史课程也得到恢复，并受到一定的鼓励。…… 随着建筑系研究生数目的日渐增加，面对年轻人迫切关注世界建筑发展动态和积极学习理论的热情，我萌发了一个想法，要为研究生开设一门较为深入地了解西方近、现代建筑思潮和建筑理论动态的课程，希望通过对西方建筑思想与实践的深入学习，引导他们对我国自己的建筑以至对世界建筑发展的进程的关注与思考。这样，从 1983 年起，就建立了以西方现、当代建筑及其理论为基础的、被称为‘建筑历史与理论’的课程，它最初是选修课，后来成为全学院研究生的一门必修课。”[1]

另一方面，尽管把建筑史教学明确分为“中国建筑史”与“外国建筑史”（在我国又以“西方建筑史”为主），或许是中国建筑学教育的一个不那么令人满意甚至已经备受质疑却又一直在实行的学科划分，但是如同汪坦先生主编《建筑理论译丛》的初衷一样，罗小未先生也特别强调指出，学习西方建筑理论的最终目的是促进中国自己的建筑学发展。用汪坦先生的话来说：“希望这套丛书能起一点这样的作用 —— 启发。让我

同济大学硕士研究生课程《近现代建筑理论与历史》课堂实况

1 罗小未：《序》，载《建筑理论的多维视野》，第 9-11(9) 页。

们来开拓耕耘自己的园地吧!”[1] 罗小未先生也指出：“思考中国建筑的现代化进程，这从某种程度上表明了我们学习西方建筑理论的根本意图，即以西方经验关照自身的发展。”[2]

当然，需要进一步追问的问题也许是如何“启发”以及如何“关照自身的发展”。对西方建筑理论的译介是否会导致对这些理论的死搬硬套和食而不化？曾经在梁思成先生那里出现的套用西方建筑的观念阐释中国建筑史所带来的问题是否会继续存在？[3] 卢永毅老师主编的《建筑理论的多维视野》已经开辟“中国现代性思考”的专门主题来回应这些问题，而这也是罗小未先生以上这段话的所指。这样的尝试将在笔者编纂的《建筑学论文读本》中得到继续。不同的是，由于这部《读本》并不刻意以主题来划分所选论文的不同范畴，因此涉及“中国问题”的论文也散落在或者说融合在整个《读本》的内容之中。在这里，所谓“中国问题”既应该包括中国学者对“近代中国”的历史研究，也不能脱离我们对当下中国和世界建筑学问题的认识。笔者相信，从世界看中国（而不是相反）（周有光先生语），进而将“中国问题”融入世界，为世界文明和人类的共同价值（其中也包括建筑学的价值）做出中国应有的贡献，这才是“外国建筑史”和“西方建筑理论”教学的真正意义。

1 汪坦：《前言》，载《建筑理论译丛》，参见曼弗雷多·塔夫里《建筑学的理论和历史》（郑时龄译），第6页。

2 罗小未：《序》，第11页。

3 关于这个问题的部分讨论，参见赖德霖：《梁思成、林徽因中国建筑史写作表微》，载《中国近代建筑史研究》，北京：清华大学出版社，2007年版，第313-330页。

2

从主线历史走向多元历史之后的思考

——威廉 · 柯蒂斯《20 世纪世界建筑史》书评[1]

1 最初以《从主线历史走向多元历史——威廉 · 柯蒂斯〈20 世纪世界建筑史〉书评》为题发表于《建筑学报》2012 年第 10 期（总第 530 期）。收入本文集时有修改。

自从黑格尔的历史哲学以及对艺术史认知有着深远影响的《美学讲演录》（*Vorlesungen über Ästhetik*，中文版《美学》诞生以来，[1]“历史主线”（main line of history）的观念一直就是艺术史学界挥之不去的幽灵。这种情况在现代建筑史学尤其是早期现代建筑史学中也不少见。从埃米尔·考夫曼（Emil Kaufmann）的《从勒杜到勒·柯布西耶：自主性建筑的起源与发展》（*Von Ledoux bis Le Corbusier: Ursprung und Entwicklung der autonomen Architektur*）到尼古拉斯·佩夫斯纳（Nikolaus Pevsner）的《现代设计的先驱——从威廉·莫里斯到瓦尔特·格罗皮乌斯》（*Pioneers of the Modern Movement: from William Morris to Walter Gropius*）[2]再到希格弗里德·吉迪恩（Sigfried Giedion）的《空间、时间与建筑》（*Space, Time and Architecture*），早期现代建筑史学的经典著作似乎都对“历史主线”情有独钟。

黑格尔 *Vorlesungen über Ästhetik* 中文版（朱光潜译）

如果说考夫曼和佩夫斯纳的著作都只是简史，因而对现代建筑史料进行去繁从简的“主线式”梳理还情有可原的话，那么吉迪恩的巨著仍然致力于勾勒一条现代建筑的“历史主线”则不能不说是黑格尔主义对现代建筑史学影响的强烈体现。作为艺术史学家沃尔夫林（Heinrich Wölfflin）的弟子，吉迪恩将黑格尔主义与沃尔夫林称之为“具有总体

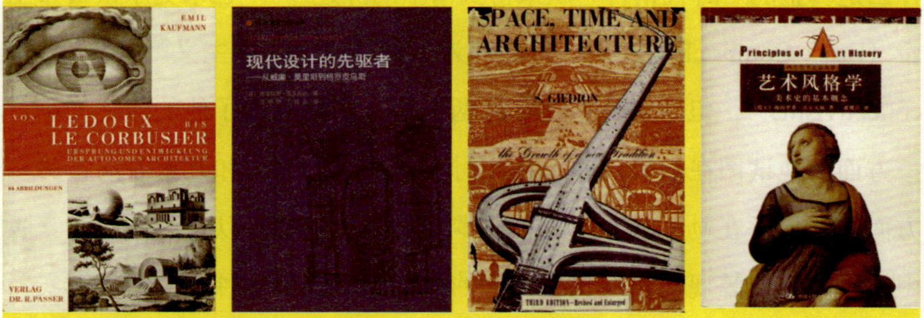

左：考夫曼《从勒杜到勒·柯布西耶——自主性建筑的起源与发展》德文版
中左：佩夫斯纳《现代设计的先驱者——从威廉·莫里斯到格罗皮乌斯》中文版
中右：吉迪恩《空间、时间与建筑》英文版　右：沃尔夫林 *Principles of Art History* 中文版

1 艺术史学家恩斯特·贡布里希曾经将黑格尔及其《美学讲演录》称为“艺术史之父”（the father of the history of art）。见 Ernst Gombrich，“Hegel and Art History,”*On the Methodology of Architectural History*, Architectural Profile 1981, pp.3-9. 中文译文见《“艺术史之父”——读 G.W.F. 黑格尔（1770-1831）的〈美学讲演录〉》”（曹意强译），载《艺术史的视野——图像研究的理论、方法与意义》（曹意强、麦克尔·波德罗主编），杭州：中国美术学院出版社，2007 年版，第 175-188 页。

2 1936年首版书名，后更改为Pioneers of Modern Design: From William Morris to Walter Gropius，中文版见尼古拉斯·佩夫斯纳：《现代设计的先驱者——从威廉·莫里斯到格罗皮乌斯》（王申祜、王晓京译），北京：中国建筑工业出版社，2004 年版。

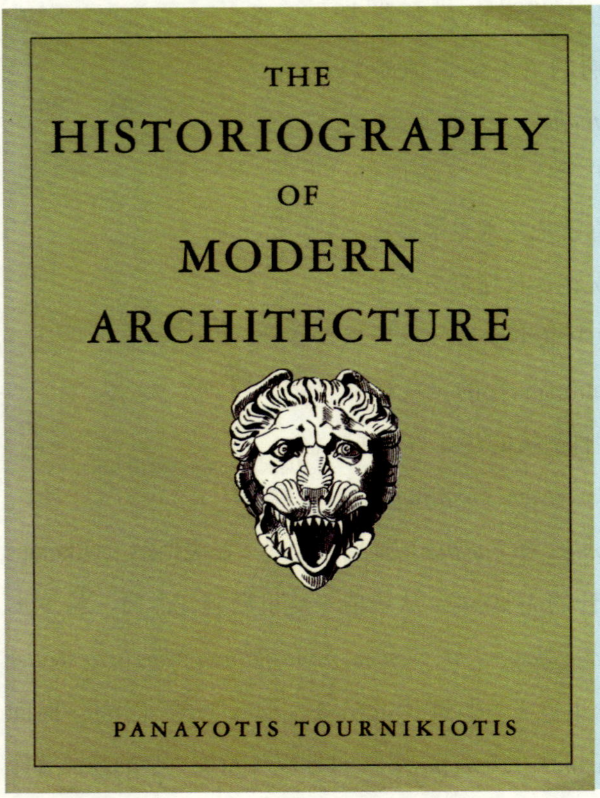

图尼基奥迪斯《现代建筑史学》英文版

代表性的形式”（the most general representational forms）的风格史观念结合起来，不仅试图勾勒符合“时代精神”（Zeitgeist）的表现形式，而且以此为现代建筑寻求历史合法性。吉迪恩在《空间、时间与建筑》的序言中这样写道：“作为艺术史学家，我是海因里希·沃尔夫林的弟子。通过与他的个人接触以及他的杰出讲座，我们——他的门生——学会了把握时代精神（learned to grasp the spirit of an epoch）。”[1]尽管什么是真正的“时代精神”从未得到清晰的陈述——艺术史学家贡布里希（Ernst Gombrich）曾经将其辨析为一种“历史集体主义”（historical collectivism）[2]，但是吉迪恩还是在两个基本层面进行了他的历史勾勒：一方面，他试图在一般层面上指出时代的技术、道德和艺术方向；另一方面他又似乎在勒·柯布西耶、格罗皮乌斯、赖特、以及后来在第三版中增补的密斯等现代建筑代表人物的作品中看到了这种方向的“真实”表达。同样，作为贯穿全书的思想红线，吉迪恩不仅将他所谓的“空间-时间”（Space-Time）的观念视为时代精神的抽象概括，而且还通过毕加索的格尔尼卡、爱因斯坦的相对论，或者从19世纪开始渐入佳境的钢结构、现代运动时期的包豪斯校舍、甚至在更广泛的城市层面的高速公

1 Sigfried Giedion, *Space, Time and Architecture: The Growth of a New Tradition* (Cambridge, Massachusetts: Harvard University Press, 1982 edition, fourteenth printing, 2002),p.2.

2 在贡布里希看来，黑格尔对艺术史的深远影响主要体现在三个方面：审美超眼主义（aesthetic transcendentalism）、历史集体主义（historical collectivism）和历史决定论（historical determinism）。见《“艺术史之父”——读G.W.F.黑格尔（1770-1831）的〈美学讲演录〉》，第176页。

左上：贝纳沃诺《现代建筑史》英文版　中上：班纳姆《大师的时代——关于现代建筑的个人见解》英文版
右上：赛维《现代建筑语言》中文版　左下：柯林斯《现代建筑思想的演变》英文版
中下：塔夫里《现代建筑》中文版　右下：塔夫里《设计与乌托邦》意大利文版

路对之进行解释。与此同时，吉迪恩也在有意无意之中履行一种排他主义的叙事模式，将表现主义（Expressionism）、装饰艺术（Art Deco）等“旁枝末节”或者像特拉尼这样的“麻烦人物”（他曾经与意大利法西斯有染）排除在现代建筑历史之外。此外，在关于“现代建筑四大师”的阐述方面，吉迪恩强调的往往是他们在现代建筑意识形态方面的共同之处以及建筑作品上的新，而大师们之间的差异——事实上这些差异是巨大的和至关重要的——以及与传统千丝万缕的关系则往往被忽略。

相较于这种建立在黑格尔主义思想模式基础之上的早期现代建筑史学，20 世纪下半叶出现的现代建筑史学著作视野更为开阔，对现代建筑的认识也更加宽容和多元。在这样的过程中，晚近的现代建筑史学也重新定位了自己与现代建筑发展的关系。用希腊学者潘纳尤迪斯·图尼基奥迪斯（Panayotis Tournikiotis）在《现代建筑史学》（*The Historigraphy of Modern Architecture*）中的观点来看，如果说早期现代建筑史学更多地呈现出一种“导向性”（operative）特征——通过现代建筑史学的建立为某种特定的现代建筑摇旗呐喊，那么这一特征经由布鲁诺·赛维（Bruno Zevi）和列奥那多·本奈沃洛（Leonardo Benevolo）的传承，逐步演变为以雷纳·班纳姆（Reyner Banham）、彼得·柯林斯（Peter Collins）和曼弗雷德·塔夫里（Manfredo Tafuri）为代表的“质疑性”（derogative）立场，尽管在图尼基奥迪斯看来，班纳姆和柯林斯的“质疑”本质上对现代建筑还是肯定的，而塔夫里则试图从根本上瓦解现代建筑的意识形态基础。[1]

值得注意的是，在这部完成于 1990 年代的现代建筑史学研究的论著中，图尼基奥迪斯对 20 世纪的现代建筑史学著作进行了归类和筛选，不仅排除了早期现代建筑史学中建筑师撰写的那部分内容 —— 因为它们形成的是一种“建筑话语”(architectural discourse)而非“历史话语”(histocial discourse)[2]，而且也忽略了文森特·斯卡利(Vincent Scully)、柯林·罗(Collin Rowe)、阿兰·科洪（Alan Coulqhon）、肯尼斯·弗兰姆普敦（Kenneth Frampton）等建筑史学家和理论家的著作 —— 因为它们“并没有真正独到的史学方法（démarche）”，要么只是图氏已经论述的几种趋势的延伸，要么虽然准则不同，但“其实还是知识的传播，而没有形成别具一格的关于当代建筑晚近史的话语”。[3]

看起来，虽然图尼奥提斯已经充分意识到吉迪恩等早期现代建筑史学中的“历史主线”“时代精神”“整体模式”等黑格尔主义的影响，但是他自己同时也似乎在不知不觉中落入了另一种黑格尔主义的陷阱 —— 在那里，“共性”（common denominator）和“类型”（typology）[4]远比差异性重要，更值得“史学史”（history of history）关注。也许，没有什么能够比这更能说明，对于艺术和建筑史学而言（当然这个问题对于艺术和建筑理论家和评论家们而言也并非毫不相干），要真正摆脱黑格尔主义 —— 或者贡布里希揭示的存在于黑格尔主义中的“历史集体主义”—— 的影响并非易事。

出于同样的原因，图尼奥提斯也将威廉·柯蒂斯（William J. R. Curtis）的《1900 年以来的现代建筑》（*Modern Architecture Since 1900*）排除在他的现代建筑史学史论述之外。但是，在该书的中文译本《20 世

左：文森特·斯卡利《现代建筑》英文版　左中：柯林·罗《美好意愿的建筑》英文版
右中：阿兰·科洪《现代建筑》英文版　右：肯尼斯·弗兰姆普敦《现代建筑——一部批判的历史》中文版

1 Panayotis Tournikiotis,“Introduction,”*The Historiography of Modern Architecture* (Cambridge, Massachusetts and London: The MIT Press, 1999), pp.14-15.

2 Ibid, p.7.

3 Ibid, p.232.

4 Ibid, p.3-4.

威廉·柯蒂斯《1900年以来的现代建筑》中文版

纪世界建筑史》问世之时，如果我们要在现代建筑史学的语境中阅读和理解这部著作，而又避免重蹈黑格尔主义的覆辙，那么一个有效的办法也许就是去关注这部著作的独特性以及与其他现代建筑史学著作的差异。

事实上，这部著作的特点相当明显。首先，从篇幅上讲，它是笔者有限的知识范围之内所接触到的篇幅最为庞大的一部现代建筑史学著作。也许，篇幅的庞大已经在一定意义上彰显了本书作者的反黑格尔主义立场。与吉迪恩力图勾画的“历史主线”不同，柯蒂斯关注的正是现代建筑的多样和复杂，并且由于这些多样和复杂并非三言两语就能简而言之，细节、案例、地区和人物之间的差异、矛盾和冲突都有至关重要的意义，都需要适当的空间展开论述。笔者以为，这应该是该书篇幅冗长的最直接原因。

说到“鸿篇巨著”，人们或许会联想到一个与黑格尔主义不无关联的概念——“宏大叙事”（grand narrative）。然而，所谓“宏大叙事”与其在字面上理解为鸿篇巨著，毋宁说是一种整体、统一、线性和决定论的历史观。因此，一篇5000字的小论文可能是“宏大叙事”，而一部洋洋百万字的巨著却有可能不是。应该说，柯蒂斯的这部著作属于后者。这是因为，尽管它实属鸿篇巨著，却把克服现代建筑史学中的黑格尔主义视为己任，不遗余力地展现现代建筑曲折多样和纷繁复杂的历史图景。为此，该书不仅涵盖了“表现主义”“装饰艺术”、特拉尼以及更多曾经在早期现代建筑史学中缺失的内容，不仅在论述现代建筑超越传统的同时也充分认识到它对传统的继承（这些都已经为吉迪恩之后的现代建筑史学广泛实践），而且还开辟专章讨论“早先传统的延续”（The Continuity of Older Traditions，第17章）、“现代建筑运动的极权主义批判”（Totalitarian Critiques of the Modern Movement，第 20 章，类似的篇章也曾出现在本奈沃洛和弗兰姆普敦的著作中）等问题。

与大多数现代建筑史学不同，柯蒂斯还在第 17 章中力求直面“折衷主义”（eclecticism）的问题，甚至可以说是为其辩护。在早期现代建筑史学中，“折衷主义”几乎臭名昭著。这或许情有可原，因为一旦“折衷”，建筑“就不再符合预先设定的历史图像，就不得不被认为有悖于时代。…… 为定义统一的现代主义，人们必须煞费苦心，要么将某一风格锁定在某一时代，要么将某一时期与某一种风格捆绑在一起”[1]。但是，在柯蒂斯看来，“风格”并不能真正决定一个建筑的品质，问题的关键在于如何“折衷”或“折衷”什么。有趣的是，柯林 · 罗曾经将勒 · 柯布西耶称为“一位兼收并蓄又创造力非凡的折衷主义者”（the most catholic and ingenious eclectics）[2]。同样，建筑史学家迪米特力 · 波菲里奥斯（Demetri Porphyrios）也曾用“现代折衷主义的源泉”（the source of modern eclecticism）来概括他对阿尔瓦 · 阿尔托的研究。[3] 显然，在这些史学家看来，风格并不重要；兼收并蓄，以现代的眼光向人类建筑文化的一切资源学习，并在这种学习中进行非凡的创造才是“折衷主义”的价值所在。

柯蒂斯著作的另一个特点是注重细节。以勒 · 柯布西耶研究为例：尽管现代建筑史学中从来不乏有关柯布的论述，柯蒂斯还是在这部著作中为我们提供了相当多之前未曾出现的细节和轶事并将它们融合在历史讨论之中，如第四章对 Dom-ino 的说明以及与佩雷和艾纳比克体系进行的比较、第十章对柯布名字的说明、第 85 页柯布用在佩雷事务所获得的第一份薪水买了维奥莱 - 勒 - 迪克（Viollet-le-Duc）的《建筑词典》（*Dictionnaire de l'architecture*），等等。一定程度上，这应该归功于该书的宏大篇幅所提供的文字空间，但也可以说明柯蒂斯这部著作的另一个关键之处，即对具体问题和事物的关注远远超过对条条框框或者各种“主义”的重视。相比之下，该书指出柯布名著 *Vers une architecture*（《走向一种建筑》）被 1927 年的英译本“误译”（mistranslated）为 *Towards a New Architecture*（《走向新建筑》）的论断则更具理论和史学意义。[4] 当然，人们也不得不看到，虽然自 1982 年首版以来已经历经 3 版和近 10 次印刷，该书仍然有一些令人惊讶的细节错误，如把 Raumplan 写成 raumplan，而且 Raumplan 还没有在索引中出现，似乎它是现代建筑发展史上一个无关

1 威廉 J. R. 柯蒂斯：《20 世纪世界建筑史》（本书翻译委员会译），北京：中国建筑工业出版社，2012 年版，第 289 页。
2 Colin Rowe, “The Mathematics of the Ideal Villa,” *The Mathematics of the Ideal Villa and Other Essays* (Cambridge, Massachusetts and London, 1976)，p.15.
3 Demetri Porphyrios, *Sources of Modern Eclecticism* (London: Academy Editions/St. Martin's Press, 1982).
4 关于这个问题的进一步论述，参见本文集《勒 · 柯布西耶 *Vers une architecture* 译名考》。

紧要的概念。《装饰与罪恶》被一带而过，相应地，该书涉及路斯的内容也十分简单，与弗兰姆普敦或者科洪在相关问题上的论述实在不可同日而语。

但是，柯蒂斯这部著作最大的不同凡响之处也许还在于它对“西方中心主义”的突破。从考夫曼、佩夫斯纳和吉迪恩开始，“现代建筑”似乎就成为西方建筑的专利。这种情况在吉迪恩之后继续延续，其中最值得一提的也许是塔夫里与达尔·科（Francesco Dal Co）合著的现代建筑史论著所遭遇的情况。这部著作 1976 年以意大利文问世，1979 年被译为英文出版，书名也由原来的 *Architettura contemporanea* 变成比较标准的 *Modern Architecture*。然而无论是意大利文版本还是英文版本，该书从一开始就是作为意大利著名结构工程师奈尔维（Pier Luigi Nervi）主编的《世界建筑史》（*History of World Architecture*）系列丛书之一来看待的（为配合在北京召开的第 20 届国际建筑师协会大会，中国建筑工业出版社将该系列丛书全套引进，于 1999 年以中文翻译出版）。有趣的是，尽管塔夫里和达尔·科的这部著作基本没有涉及欧美之外的建筑（只是简单提及日本和巴西建筑），但是它在整个系列丛书中却占据了一个与从古希腊到 19 世纪的西方建筑相当的历史地位。换言之，在这个系列丛书的编者和出版者看来，“现代建筑”与“希腊建筑”“罗马建筑”“哥特建筑”“文艺复兴建筑”或者“巴洛克建筑”一样，可以等同于西方建筑，只有“远古建筑”“东方建筑”和“伊斯兰建筑”是例外。这或许可以解释天津大学邹德侬先生在翻译本奈沃洛的《现代建筑史》（*History of Modern Architecture*，意大利版本原名为 *Storia dell'architettura moderna*）时将中文译本更名为《西方现代建筑史》的原因——在邹先生那里，这一点似乎如此不言自明，以至于他在中译本序言中对书名的变化没有予以任何解释。[1]

列奥纳多·本奈沃洛《现代建筑史》中文版

诚然，“西方”从来都不是一个纯粹地理意义上的概念（如地理意义上的欧洲或北美），而是充满地缘政治性，比如人们会将日本视为“西方集团”的一部分。同样，澳大利亚也常常理所当然地被视为“西方国家”，而

1 见 L·本奈沃洛：《西方现代建筑史》（邹德侬、巴竹师、高军译），天津科学技术出版社，1996。

真正在地理上位于欧洲的土耳其却因其伊斯兰文化而难以成为“西方”的一部分。在本奈沃洛的著作中，以尼迈耶为代表的巴西这类在传统地缘政治意义上不属于“西方”国家的建筑被置于“新的国际领域”的范畴之中。[1]这种“突破”或多或少也曾出现在其他现代建筑史学家的著作之中，而弗兰姆普敦的《现代建筑：一部批判的历史》则在1985年和1992年的修订版中分别以“批判地域主义”（critical regionalism）和“世界建筑”（world architecture）的名义相当有限地提到了墨西哥的巴拉干（Luis Barragán）、印度的柯里亚（Charles Correa）等“发展中国家”的建筑师和他们的作品。按照弗兰姆普敦的说法，“批判地域主义”的要旨之一就是反对“中心主义”（不仅是“西方中心主义”，因为即使在“西方”也有“中心”和“边缘”的对立）。[2]但是，在笔者看来，弗兰姆普敦的“批判地域主义”其实并未真正能够化解“中心霸权”；相反，从“中心”俯视“边缘”恰恰是弗兰姆普敦式的“反中心主义”根深蒂固的内在矛盾。这无疑是一个十分复杂的问题，并且鉴于当代世界政治、经济和文化格局，也将不可避免地存在于一切西方的“反西方中心主义”（包括柯蒂斯的著作）之中。尽管如此，我们仍然可以说，像柯蒂斯那样开辟专门章节论述“发展中国家”或者“非西方国家”（见该书第21、27、31和34章）实属现代建筑史学之罕见。至少，就其涉及“非西方”国家和地区的广度和篇幅而言，它或许是第一部无需更名为“西方现代建筑史”的现代建筑史。有趣的是，柯蒂斯这部著作的书名直译应该是《1900年以来的

上：北京建设部大楼　下：台北中正纪念堂

右页
左上：查尔斯 · 柯里亚　甘地纪念馆
右上：巴尔什科纳 · 多西　工作室
左下：尤恩 · 伍重：科威特议会大厦
右下：杰弗里 · 巴瓦　坎达拉马遗产酒店

1 L · 本奈沃洛：《西方现代建筑史》，第20页。
2 肯尼斯 · 弗兰姆普敦：《现代建筑：一部批判的历史》（张钦楠等译），北京：生活 · 读书 · 新知三联书店，2004年版，第355页。

现代建筑》，中文版将其更名为《20 世纪世界建筑史》本来是要以一个更为响亮和宏大的标题申请国家出版基金，现在看起来倒更像是一个歪打正着的举措。

柯蒂斯所谓的“发展中国家”也包括中国。事实上，正是在这个题为《发展中国家的现代性、传统与文化认同》(“Modernity, Tradition and Identity in the Developing World”) 的第 31 章中，中国的 20 世纪建筑首次出现在一部有重要影响的现代建筑史学著作之中，尽管与全书的鸿篇巨幅相比，这些内容只能算是寥寥无几。不幸的是，即便是这寥若晨星的几笔 —— 它们列举了北京的人民大会堂、历史博物馆、建设部大楼（图 18）以及台北的中正纪念堂等建筑，柯蒂斯的总体评价也是否定性的，谓之“对东方传统的触碰依然是表面化的”（its Oriental touches were still skin-deep）[1]。这是一个可以引发许多问题，甚至令人产生无限遐想的有趣现象。比如，人们也许可以质疑，柯蒂斯是否有口是心非之嫌，一方面大讲多元的史学观和反西方中心主义，反对唯现代风格论，另一方面却在骨子里无法真正认同非西方的“他者”。但是，如果我们注意一下柯蒂斯在同一章节以及其他章节论述发展中国家或非西方国家的“地域主义”的建筑作品时使用的肯定性甚至是赞誉有加的语言（当然也有批评的地方），我们便很难用“口是心非”来责难柯蒂斯。人们或许还可以认为柯蒂斯是出于政治意识形态的原因而对中国 20 世纪五六十年代的建筑持否定态度的。但是，如果我们注意到台北的中正纪念堂同样在柯

1 柯蒂斯：《20 世纪世界建筑史》，第 582 页。英文原文见 William J. R. Curtis, *Modern Architecture since 1900* (London and New York: Phaidon, Third Edition, 1996), p.582.

杨廷宝
左上：京奉铁路沈阳总站
左中：自宅
左下：中央医院（现南京军区总医院）
右上：清华大学图书馆扩建
右中：北京和平饭店
右下：国民党中央党史史料陈列馆（现中国第二历史档案馆）

蒂斯的批评之列，我们便很难相信政治意识形态是柯蒂斯评判的主要标准。人们或许还可以认为，柯蒂斯并不真正了解20世纪中国现代建筑，他也许既对中国曾经拥有杨廷宝这样的建筑师及其最优秀的作品一无所知——在笔者看来，20世纪中国建筑中或许没有什么比杨廷宝的建筑能够更好地诠释柯蒂斯主张的真正好的建筑品质是超越风格之上的观点，但也没有什么能够比杨廷宝的建筑更能说明仅仅是历史风格复兴既不能真正推动中国建筑学的现代发展，也不能真正发扬光大中国传统建筑文化这一点；也对“营造学社”的学者们在挖掘和研究中国建筑文化遗产方面的艰苦卓绝的努力和成就闻所未闻。即便在1949年之后的中国建筑之中，他似乎也没有意识到还有以夏昌世为代表的岭南学派，或者以冯纪忠为代表的沪上学派在中国建筑之“现代性、传统与文化认同”方面进行的与上述官方建筑全然不同的探索，而这些探索完全可以为他的论述提供更好的案例，更不要说王澍、大舍、TAO迹等新一代中国建筑师在这方面的思考和追求，尽管这些都是一些更新的案例，完全可以用21世纪建筑在柯蒂斯这部著作中的整体缺位来进行解释。

上：夏昌世 中山医学院教学楼
下：冯纪忠 何陋轩

但是，如果我们果真这样想了，

左上：王澍 陆文宇　中国美术学院建筑与环境艺术学院　　右上：王澍 陆文宇　文村
左下：大舍　上海龙美术馆　　右下：TAO 迹　武夷竹筏厂

我们便已经涉及一个与柯蒂斯的立场不无相关的更为有趣、也更为重要的问题。这个问题可以这样表述：是什么使我们认为后面的这些案例要比柯蒂斯批评的那种对传统的“表面化触碰”更值得赞许？在这些赞许之后，有没有我们仍然需要坚持的现代建筑基本价值（而又无需重蹈以CIAM为代表的“现代建筑运动”的意识形态谬误或者“国际风格”的形式教条[1]）在发挥作用？在更为广泛的层面上，“地域”和“多元”是否可以成为一套脱离这些基本价值的自我话语的理由和借口？

没错，柯蒂斯的著作表明，他的史学立场与吉迪恩式的“历史主线”意识背道而驰，他反对“将历史比拟为人生”，反对“建筑发展也要经历出生、青年、成年以及最终死亡”的线性历史观，[2] 力图用“地域”和“多元”来取代“历史主线”和“决定论”。但是，“多元主义”的问题往往也在于，它容易缺乏批判性，从而陷入相对主义（relativism），似乎“多元”就是毫无标准，怎么都行，“地域”就是自说自话，没有任何共同价值标准可言。一定意义上，正如建筑史学家阿兰·科洪（Alan Colquhoun）曾经指出的，相对主义也是一种历史主义（historicism），它认为“一切社会文化现象都是历史决定的，所有真理都是相对的”[3]。而在黑格尔那里，按照贡布里希的分析，相对主义则是“形而上乐观主义”（metaphysical

---

1 关于这些问题的讨论，见 Colin St John Wilson, *The Other Tradition of Modern Architecture: The Uncompleted Project* (London: Black Dog Architecture, Revised Edition, 2007).

2 柯蒂斯：《20 世纪世界建筑史》，第 21 页。

3 Alan Colquhoun, "Three kinds of Historicism," *Modernity and the Classical Tradition: Architectural Essays 1980-1987* (Cambridge, Massachusetts and London, The MIT Press, 1989), pp.3-19(3).

optimism）和辩证法（the dialectic）的产物，是一种"辩证相对主义"（dialectic relativism），它将一切艺术的价值视为相对的，因为"'艺术绝不是精神表现的最高形式'。它为沉思所瓦解，为纯思想、为哲学所取代。由此艺术属于过去"。[1]

如果说后一种相对主义与我们这里的讨论没有太多相关性的话，那么值得注意的是柯蒂斯对前一种相对主义的警觉。早在一篇关于现代建筑史学著作的书评中柯蒂斯就已经指出：

> 如果认为，只要驱逐了时代精神的幽灵便万事大吉，这样的想法实在是过于天真。事实上，另一些幽灵已经悄然而至。20 世纪的史学家不得不在一种和之前一样扭曲的神话中审视他的主题。这类神话大多源自相对主义（relativism）、民粹主义（populism）以及某些可以称之为"对习俗的顶礼膜拜"（the "cult of the convention"）的东西。[2]

柯蒂斯说这番话时似乎没有明确针对某个特定的作者或者著作，但是在笔者看来，如果在 20 世纪后半叶完成的建筑史学著作中有什么能够说明，一旦丧失批判性便有可能从"多元"迅速滑向彻底的"相对主义"，那么没有其他什么比新版《弗莱彻建筑史》更能够成为一个典型案例。这部曾以《比较方法的建筑史》（*History of Architecture on the Comparative Method*）命名、却以第 4 至第 16 版中的一棵"建筑之树"而使数代中国建筑学者耿耿于怀近一个世纪之久的史学著作，从第 17 版开始逐步放弃所谓"历史性风格"（the historical styles，即正统的欧洲建筑文化主线）和"非历史性风格"（the non-historical styles，即正统的欧洲建筑文化主线之外的建筑，特别是非欧洲文化的建筑）的区分以及因这一区分而饱受诟病的"欧洲中心主义"（如果不是黑格尔意义上的历史主线的话），转而演变为一种大百科全书式的建筑史汇编，并在 1996 年问世的第 20 版中增加了中东建筑、东南亚建筑、印度次大陆建筑、东欧建筑、拉丁美洲建筑等章节。此外，土耳其学者纳尔班托格鲁（G. B. Nalbantoglu）曾经指出："……更多民居建筑，更多工程结构，更多当代设计：如果不是因为缺少更多女性建筑师的话，20 版的弗莱彻建筑史真可以被视为对 20 世纪晚期关注点最为应时的修改。"[3] 也是在这一版中，

---

1 贡布里希：《"艺术史之父"——读 G.W.F. 黑格尔（1770-1831）的〈美学讲演录〉》，第 178 页。

2 William Curtis, "Manfredo Tafuri and Francesco Dal Co, *Modern Architecture* (New York: H. N. Abrams Inc., 1979, 448 pp., 673 illus. $42.50); Kenneth Frampton, *Modern Architecture: A Critical History* (New York and Toronto: Oxford University Press, 1980), 324 pp., 297 illus. $17.95 (cloth); $9.95 (paperback)," Journal of the Society of Architectural Historians, XL:2, May 1981, pp.168-170(168).

3 Gülsüm Baydar Nalbantoglu, "Toward Postcolonial Openings: Rereading Sir Banister Fletcher's History of Architecture", *Assemblage* 35, 1998, pp.6-17(8).

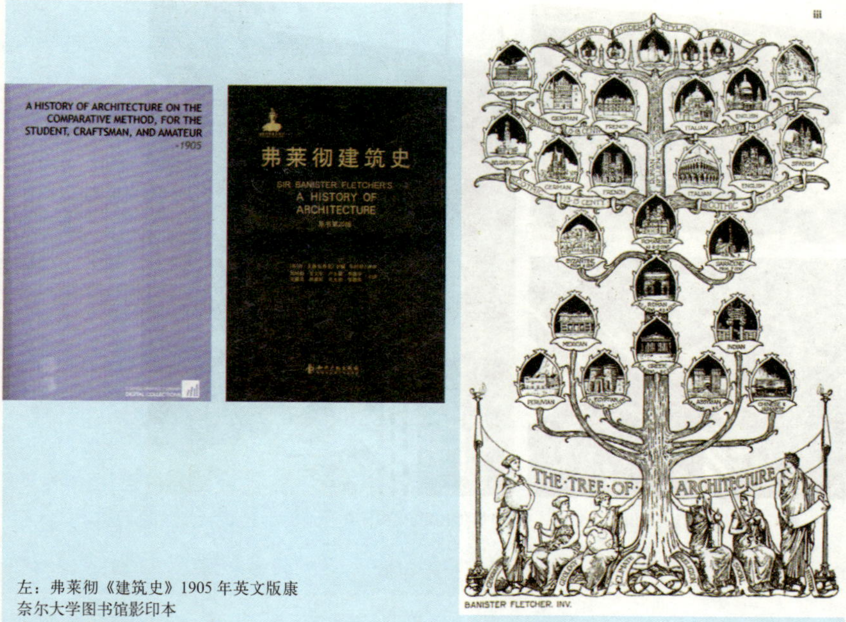

左：弗莱彻《建筑史》1905 年英文版康奈尔大学图书馆影印本
中：《弗莱彻建筑史》中文版
右："建筑之树"

20 世纪中国建筑成为单独的一章（第 53 章）。遗憾的是，除了史料的苍白罗列（暂且不说史料的选择本身是否值得商榷），人们在这一章中看不到任何具有批判意义的立场和论述。看起来，在其编者和作者那里，批判性已经没有必要，因为正如罗小未先生指出的（尽管罗先生对之是完全赞许的），该书的思想基础是"相对主义"。[1]

因此，柯蒂斯的这部现代建筑史学著作的价值不仅在于它以反对早期现代建筑史学的"历史主线"和"决定论"为己任，倡导"多元"和"地域"的历史观，而且更在于它在这样做的时候，也提出了谨防相对主义的必要性和迫切性。具体来说，通过对"多元"和"地域"作品的批判性评述（这样说并不意味着我们必须完全赞同柯蒂斯的观点），柯蒂斯这部现代建筑史学著作更为重要的意义在于让我们在进入 20 世纪以后的世界建筑语境中思考这样的问题：什么是北京人民大会堂、历史博物馆、建设部大楼以及台北的中正纪念堂等建筑在他的眼中不足为取的原因？或者以更具当下意义的建筑为例，为什么像 2010 年上海世博会中国馆这一被赋予中国式政治和文化意义且得到官方和民众普遍认可的建筑却难以在世界建筑学之中获得一个可以与其在中国的声望相匹配的地位？相比之下，是什么使得王澍建筑得到世界建筑学的普遍认可（尽管这一点并不意味着王澍的建筑完美无缺或者我们应该忽视对它的批评）？什么是王澍建筑中的"多元性"和"地域性"（当然，这里还有一个本文有

1 罗小未：《中文版序》，载《弗莱彻建筑史》（丹·克鲁克香克主编，郑时龄译审，郑时龄、支文军、卢永毅、李德华、吴骥良、郭黛姮、吴光祖、邹德侬主译），北京：知识产权出版社，2011 年版，第 i-iv(iv) 页。

左：施工中的上海世博会中国馆　右：上海世博会中国馆结构模型

限的篇幅不容展开讨论的“个体性”问题[1]），什么又是其中的“世界性”和“普遍性”？我们需要的是否果真如王澍所言在于用“中国建筑学”来抗衡“西方建筑学”？我们究竟应该在什么意义上理解柯蒂斯所说的，“……设想一种‘普遍性’（to posit a ‘universalizing’ aspect）并非完全没有道理，前提是人们能够摆脱这一立场背后的西方偏见（the Western bias）和进步假设（progressive assumptions）”。[2]

在笔者看来，对于当代中国建筑学的发展，这些问题并非完全没有意义，而且说到底，它们所涉及的也绝非仅仅是一个“话语权”问题。就此而言，尽管柯蒂斯对20世纪中国建筑的论述少之又少——事实上，如果没有这部分少得可怜的内容，我们完全可以将该书的书名译为《1900年以来的外国现代建筑》，但是这种不足并不妨碍它仍然是一部发人深省且特别具有中国意义的现代建筑史学著作。

1 在该书的“结论”部分，柯蒂斯这样总结道：“本质上，现代主义建筑和历史上的建筑相比并无高低之分，好坏的唯一标准取决于对具体情况恰当地加以使用。在创造品质的问题上没有任何捷径，也没有万能灵药。以时髦新鲜作为目标的建筑，其价值也和时尚一样转瞬即逝。事实上，我们不必太在乎‘现代性’，因为真正重要的是真实性。”在此，柯蒂斯不仅对他所谓的“真实性”进行了阐述，而且还特别强调了个体的重要性，因为“被脱离了真实的个人风格的基础，上述真实性是无法想象的。”（见《20世纪世界建筑史》第689页。）可以认为，也正是这种对“个体性”的强调在一定意义上使柯蒂斯得以超越“主义”和“时代精神”的桎梏，将一个更为细微和多样的现代建筑发展图景呈现在读者面前。

2 柯蒂斯：《20世纪世界建筑史》，第372页。

3

From Mainline History to Plural History, and So What?[1]

1 Originally presented as a keynote speech at the Third Symposium of the Forum of Contemporary Architectural Theories organized by AA School of Architecture, Southeast University and Shanghai Xian Dai Architectural Design (Group) Co.,Ltd, and held at Southeast University in Nanjing, China, September 15-16, 2012.

Ever since the posthumous publication of Hegel's *Vorlesungen über die Ästhetik*, which, according to Ernst Gombrich, "should be regarded as the founding document of the modern study of art"[1], the obsession with mainline history has haunted the field of art history, and this is no less so when it comes to the historical writing on modern architecture. From Emil Kaufmann's *Von Ledoux bis Le Corbusier: Ursprung und Entwicklung der autonomen Architektur* to Nikolaus Pevsner's *Pioneers of the Modern Movement: from William Morris to Walter Gropius* (or *Pioneers of Modern Design* as it was named later on), and Sigfried Giedion's *Space, Time and Architecture*, the early milestones of the history of modern architecture persisted in one way or another with a mode of thinking that served to shape so much of the mainline history of modern architecture.

If the mainline habit of mind that one finds in Kaufmann's and Pevsner's aforementioned works could possibly be excused for their concise format, Giedion's *magnum opus*, which no less dedicated itself to a mainline concept of history, provided only another testimony to the impact of Hegelianism on the historical writing on modern architecture. As a student of Heinrich Wölfflin, Giedion combined Hegelianism with Wölfflin's notion of "the most general representational forms" in his articulation of modern architecture that could be understood as a coherent expression of *Zeitgeist*, or the spirit of time, which in turn was to ensure the historical legitimacy of modern architecture. The introduction to *Space, Time and Architecture* reads: "As an art historian I am a disciple of Heinrich Wölfflin. In our personal contacts with him as well as through his distinguished lectures, we, his pupils, learned to grasp the spirit of epoch."[2] From the outset, without making clear what was meant by the idea of "the spirit of epoch", which according to Ernst Gombrich should be in essence treated as a "historical collectivism"[3], Giedion's historical narrative of modern architecture was marked basically in two ways: the formulation of technological, ethic and artistic directions on one hand, and validation of the work by Le Corbusier, Gropius, Frank

---

1 Ernst Gombrich, "Hegel and Art History" in *On the Methodology of Architectural History*, ed. Demetri Porphyrios, Architectural Design Profile 1981, 3-9(3).

2 Sigfried Giedion, *Space, Time and Architecture: the Growth of a New Tradition* (Cambridge, Massachusetts: Harvard University Press, 1982 by Andreas Giedion and Verena Clay-Giedion, fourteenth print 2002), p.2.

3 Ernst Gombrich, "Hegel and Art History", p.3.

Lloyd Wright (later expanded to Mies van der Rohe and Alvar Aalto, and even later to Jørn Utzon) on the other. The same is true for the highly abstract "Space-Time" which ran like a red thread throughout the book interpretated by Giedion in divergent terms, be it Picasso's *Guernica* or Einstein's *Theory of Relativity*, and in architecture and urban planning, instances ranging from the 19th century French steel construction to the Bauhau building of high modernism, or, curiously enough, urban motoways. As such, its histocial narrative not only locked out such aberrances as Expressionism and Art Deco, but also ruled out the intricate figures like Giuseppe Terragni and the Italian futurists and rationalists alike who were in one way or another entangled with the Italian Fascism. Meanwhile, as to the "Four Masters", Giedion was keen to stress their similarities, either in aesthetic, technological or ideological terms, while their disparities, which were no less significant to the formation and evolution of modern architecture, were largely ignored. The same can be said to his account of these architects by focusing on what was new in their works, whereas their connection to tradition largely fell out.

Against this background, the historical writings in the second half of the 20th century have been characterized by more nuanced and complex pictures of what happened in the evolution of modern architecture. What is more, as architectural historian Panayotis Tournikiotis suggests in his *Historiography of Modern Architecture*, divergent to the protagonist position of the early historical writings on modern architecture, the succeeding historians, Bruno Zevi, Leonardo Benevolo, Reyner Banham, Peter Collins and others, seemed to hold a more "derogative" stand that tended to valorize and disqualify modern architecture at the same time.[1] In this latter regard, Manfredo Tafuri was apparently the one who was standing out in his relentless critique of the ideology of modern architecture that ultimately led to the undermining of its very foundation – *l'architeture assassinée*, to use Aldo Rossi's words in his hand-painted plate from 1975.[2]

However, in order to identify what he calls the *démarches* of the writings on the history of modern architecture, Tournikiotis' treatise

---

1 Panayotis Tournikiotis, *The Historiography of Modern Architecture* (Cambridge, Massachusetts and London: The MIT Press, 1999), "Introduction" and in particular pp.14-15.

2 See Manfredo Tafuri, *Architecture and Utopia: Design and Capitalist Development*, trans. by Barbara Luigia Penta (Cambridge, Massachusetts, and London: The MIT Press, 1976), p.180.

not merely excludes the writings by architects because what they formulated effectively belonged to an "architectural discourse" rather than a historical one,[1] but also succeeded in ruling out the works of such architectural historians and theorists as Vincent Scully, Collin Rowe, Alan Colquhoun, Kenneth Frampton, William Curtis for the reason that, as Tournikiotis claimed, they "either extrapolate tendencies we have already discussed here or correspond to criteria ... connected more with the communication of knowledge"[2]. In doing so, Tournikiotis' historiographical study focuses primarily on "common denominator" and "typology"[3] rather than disparity and divergence between the historical writings on modern architecture. This focus, in a sense, only serves further to indicate that the impact of Hegelianism, or "historical collectivism" to reiterate Gombrich's terms, still imparts its mark on the discourse on architectural history, and in this instance the history of the history of modern architecture.

Therefore, when coming to *Modern Architecture Since 1900* by William J. R. Curtis that, first published in 1982 and got named only once in Tournikiotis' *Historiography of Modern Architecture*,[4] is now translated into Chinese, we would be better off, if we want to keep away from Hegelianism, not looking for the "common denominator" with other agents of the historical writings of modern architecture, but the difference and divergence that it carries within itself, and emphatically this is not only because it corresponds to the criteria connected to the communication of knowledge, but also, to say in terms of what Tournikiotis has denied, "with the formation of a different discourse about the history of the recent past of contemporary architecture."[5]

In fact, this is a book remarkable in a number of ways and for a number of reasons. First of all, to my knowledge, it is the most massive one consisting of more than 700 pages in total. The massiveness of the book itself might serve as an alibi to help explicate the author's ambition for both a broad and in-depth assessment of modern architecture where details, differences and even contradictions between cases, regions and figures are of significance and deserve

1 Ibid, p.7.
2 Ibid. p.232.
3 Ibid. pp.3-4.
4 Tournikiotis, ibid., p.232.
5 Ibid.

their respective space to unfold. This is no less so when it comes to the elaboration of modern architecture in Europe that normally constitutes the major substance of the history of modern architecture.

Not surprisingly, reading through the pages, one is often brought to rich historical details. In particular, compared to other survey writings of modern architecture, Curtis' work does provide exceptionally abundant information and even anecdotes that help to throw light on Le Corbusier's position in and his relationship with the evolution of modern architecture. His discussion of the Maison Dom-ino is made against the background from François Hennebique to Auguste Perret. And according to Curtis, when working for Perret, Le Corbusier used his first pay-packet to buy Viollet-le-Duc's *Dictionnaire de l'architecture*. Through a discussion on the English translation of *Vers une architecture* into *Towards a New Architecture*, furthermore, compelling issues are raised to counter the simplistic reduction that all too often gives rise to the misconceptions about Le Corbusier's seminal book. Such a detail-laden approach could be the reason for the bulky work of Curtis' kind, but far more importantly contributes to build up a unique work on the history of modern architecture in which historical narratives stem from the specific and the concrete rather than from doctrines or "isms".

On the other hand, this virtue of the book seems to be flawed to some degree by the fact that, although the book has been revised three times since its first edition in 1982, a few small misprints and inaccuracies are yet to be corrected. For instance, *Raumplan* is written as *raumplan*, and more surprisingly, it fails to get a place in the index, as if it was a trivial concept in the history of modern architecture. "Ornament and Crime" is slid over, and Adolf Loos is treated quite unsophisticatedly when compared with the insightful discussions made by Frampton and Colquhoun in their respective works.

Structurally, the book not only takes into account of the issues of "Expressionism", "Art Deco" and other divergences that were by and large overlooked by the early historians, but also offers a particular chapter entitled "Totalitarian Critiques of the Modern Movement" in attempt to deal with the inner contradiction of the Modern Movement in terms of the intricate relationship between Terragni and Fascist architecture. It is certainly not the first time that Terragni is included in a critical examination of the history of modern architecture. Before Curtis, Frampton devoted a chapter of his *Modern Architecture: a Critical History* to "Giuseppe Terragni and the architecture of Italian Rationalism". However, while Frampton chose to focus on the theme of Rationalism and thus more or less played down the issue regarding the relationship between the Italian Rationalist movement and Fascism, Curtis' narrative discloses cogently a different face of modern architecture in the context of Fascism and the totalitarian powers. At this point, only Tafuri and Francesco Dal Co's examination of the same issue in their *Modern Architecture* can be placed on a par,[1] though compared to Curtis' totalitarian critiques of modern architecture which "bit deep and revealed problems and rifts between avant-garde culture and some of the traditional, preservationist and institutional functions of monument architecture"[2], Tafuri and Dal Co's concern seemed to impart a much more miscellaneous narrative to the whole issue.

It is, therefore, not so much the range of historical materials than their examination and explication – thus a different discourse – that is remarkable in Curtis' work. To this I may add Chapter 21 entitled "International, National, Regional: the Diversity of a New Tradition", Chapter 26 "Disjunction and Continuities in the Europe of the 1950s" or Chapter 32 "Pluralism in the 1970s", if not to say Chapter 25 on "Alvar Aalto and Scandinavian Development", a theme that Giedion already noticed, or Chapter 30 "Extension and Critique in the 1960s", of which a similar discussion might be found in Frampton as well. Of particular interest, I should also mention Chapter 17 "The Continuity of Older Traditions", in which both the relationship between modern architecture and earlier traditions and the issue of "eclecticism" are examined. In this latter regard, while the early historical writings of modern architecture made "eclecticism" a nasty word referring to the reckless blending of historical styles, Curtis' notion of eclecticism or the eclectic has little to do with the style. This brings to mind Colin Rowe's characterization of Le Corbusier as "the most with catholic and ingenious eclectics"[3]. In a similar vein, architectural historian Demetri Por-

1 See Manfredo Tafuri and Francesco Dal Co, *Modern Architecture* (Milano: Electa / Rizzoli, 1976), Chapter 15.

2 William Curtis, *Modern Architecture since 1900* (London: Phaidon Press Limited, third edition, 1996), p.369.

3 Colin Rowe, "The Mathematics of the Ideal Villa" in *The Mathematics of the Ideal Villa and Other Essays* (Cambridge, Massachusetts and London: The MIT Press, 1976), pp.1-27(15).

phyrios summed up his study of Alvar Aalto by entitling his book as *Sources of Modern Eclecticism.*[1] Obviously, to Curtis, as to Rowe and Porphyrios, style is not the issue. Rather, learning from all sources of the culture of architecture with modern eyes and making ingenuous creations on that basis is the real value of "eclecticism".

However, the most extraordinary feature of Curtis' book must be its commitment to rise above the "Western-Centrism" in the context of the history of modern architecture. From Kaufmann, Pevsner and Gideon onwards and for the mainstream academia, the term "modern architecture" has seemingly become synonymous with Western architecture. Nothing is perhaps more striking in this regard than what happened to Tafuri and Dal Co's *Modern Architecture*, the English version of their original work in Italian, *Architettura contemporanea.* Although referring little to what happened outside of Europe and North America, it was included in the series of books under the headline "History of World Architecture" to take hold the concept of modern architecture, as if, for the editors of the series, "modern architecture" were the monopoly of the West in the same way as Greek, Gothic or Renaissance architecture, and only Oriental or Islamic architecture could be treated otherwise. This polemic is suggestive of why, in translating Leonardo Benovolo's *History of Modern Architecture* into Chinese, Zou Denong, a Chinese architectural historian from Tianjin University, found obliged to rename it as *History of Modern Architecture in the Western World.* To Zou, as to many others, this is what the book is really about.

To be sure, what is called "West" or "Western" is never a geographical issue alone. Rather, it is always full of geopolitical implications. Thus, Japan is sometimes admittedly classified as a part of the Western world, and likewise, in spite of its geographical location, Australia is taken for granted as a Western country. As contrast, one may refer to Turkey that, while geographically located in Europe, is hardly identified as a country that belongs to the West because of its Islamic culture. That said, it is worthwhile to note some serious efforts to extend the scope of the history of modern architecture and to break away from a center dominating mainline history. For instance, in Benevolo's *History of Modern Architecture*

1 Demetri Porphyrios, *Sources of Modern Eclecticism* (London: Academy Editions/St. Martin's Press, 1982).

first published in Italian in 1960 and thereafter in English in 1971, modern Brazilian architecture represented by Oscar Niemeyer and others is encapsulated into the category of the "new international sphere". Since then, this "breakthrough" has more or less become a standard in the writings of modern architecture historians. What is more, from the second edition of Kenneth Frampton's *Modern Architecture: a Critical History* of 1985 and onwords, architects such as Barragan from Mexico and Charles Correa from India were covered under the title of "critical regionalism". However, as some critic has pointed out, while the point of "critical regionalism" is, among others, to make resistance to "centrism", meaning not merely Western-centrism, but also in the West itself between center or centers and peripheries, Frampton's "critical regionalism" seems hardly to succeed in undermining the hegemony of "centrism', for no matter how tacit it might be, it carries within itself an innate contradiction by possessing a centrist, if not "post-colonial" as the same critic has asserted, position looking down to the peripheries.[1]

Given the political, economical and cultural conditions of the world that we are living in, what is raised by the critique of this kind is a polemic that is probably inevitable in any Western made "anti-Western-Centrism" including, I would say, the writing of Curtis. Despite that, it is a merit of the present book by Curtis in which an unprecedented space is given to architecture in "developing countries" and "non-Western countries". This is particularly legible in Chapters 31 and Chapter 34. Indeed, to the extent that the 20th century architecture beyond the Western world can have on the history of modern architecture as Curtis' book delineates, it is possible to say that for the first time one has a major work on the history of modern architecture that does not have to be renamed as the *History of Modern Architecture in the Western World* as Zou Denong did with the Chinese edition of Benevolo's book.

Now, it is in Chapter 31 "Modernity, Tradition and Identity in the Developing World" that the architectural works of 20th century China are represented. What is interesting, however, is that while such grand state buildings as the Great Hall of the People, the Museum

1 See for instance Keith Eggener, "Placing Resistance: A Critque of Critical Regionalism" in *Architectural Regionalism: Collected Writings on Place, Identity, Modernity, and Tradition* (Vincent B. Canizaro, editor, Princeton Architectural Press, 2012), pp. 395-407.

of Chinese History and the Revolution, the Building Ministry in Beijing, all built in 1950s, and the Chiang Kai-Shek Memorial Hall in Taipei are named to indicate modern Chinese architecture, a rather disapproving tone of the author's overall assessment can hardly go unnoticed, saying "their Oriental touches were still skin-deep"[1].

A number of questions may be raised here. One might possibly wonder whether, Curtis' critique of the aforementioned examples of Chinese architecture stems from his position that, while talking about plural history and anti-Western-Centrism, in reality leaves little space for the "Other". But if we take notice of his rather laudatory passages about many "regional" buildings in the developing or "non-Western" countries in the same chapter and elsewhere in the book, it would be difficult to consolidate that doubt. One could then, in line with the more official Chinese reaction to any critique to China, or, in the case of Michelangelo Antonioni's documentary on China, even to the portrait of China that is officially regarded discordant, suspect whether his disapproving attitude to the Chinese architecture is derived from a political ideology that in the end is anti-communist and anti-China. However, the fact that his critique refers to the Monument of Chiang Kai-Shek in Taiwan where an entirely different political system reigns would quickly dispel that doubt.

A more justified questioning seems to involve the uncertain spectrum of modern Chinese architecture of the 20th century that Curtis' book has provided. Most likely he has no idea about the works of Yang Tingbao, an outstanding architect whose buildings, in spite of their apparently divergent styles, may serve as the cogent evidence of Curtis' proposition that the quality of "architecture has nothing to do with the styles"[2]. He seems unaware of the fact that, apart from the official buildings that he refers to, there do exist different and more convincing approaches to "Modernity, Tradition and Identity" in Chinese architecture after 1949 to which the architecture by Xia Changshi and others known as the Lingnan School in the region covering Guangdong and Guangxi Provinces, or by Feng Jizhong, particularly his Square Pagoda Garden in the outskirt of Shanghai, might be referred to, not to

1 William Curtis, *Modern Architecture since 1900*, p.582.

2 Ibid, p.689.

say the more recent building works of Wang Shu, Deshaus, Trace Architecture Office (TAO) and other younger emerging architects that reasonably fall beyond the scope of Curtis' book ending up with the 20th century.

But we are at the hinge of an argument that, if we are indeed in favor of these latter cases than the Chinese buildings named by Curtis, then for what reason or on what ground are we able to do so? Speaking of the history of modern architecture, are there any basic values of modern architecture that we should hold to in making this judgment without falling back on the modernist fallacies that Colin St John Wilson has defied as irrelevant to "the other tradition of modern architecture"[1]? Could the "regional" and the "pluralistic" be served as an excuse to detach oneself from these values?

Once again it is a virtue of Curtis that, in striving for a plural and regional spectrum of the history of modern architecture rather than mainline history, determinism, *Zeitgeist* and all the rest, he keeps us alert to the problem with pluralism that, by locking out criticality, all too easily turns into synonymous with relativism, believing anything goes and nothing of interest is to be said about values in general. As he pointed out as early as in 1981 in a conflated book review on the historical account on modern architecture by Frampton and Tafuri: "It would be nice to think that the chasing out of the Zeitgeist devils left a clean house. Actually a whole lot of other poltergeists came in through the basement. The historian of the 20th century has to look at his subject through a refractory atmosphere of myths every bit as distorting as the earlier ones. Most of these stem from the holy ghosts of relativism, populism, and what one might call the 'cult of the convention'"[2].

Curtis did not refer to any particular author or work in this regard. But among the historical works on architecture completed in the second half of the 20th century, none is perhaps more illuminating to Curtis' statement than the latest edition of Sir Banister

1 Colin St John Wilson, *The Other Tradition of Modern Architecture: The Uncompleted Project* (London: Black Dog Architecture, Revised Edition, 2007).

2 William Curtis, Review: *Modern Architecture* by Manfredo Tafuri, Francesco Dal Co; *Modern Architecture: A Critical History* by Kenneth Frampton in *Journal of the Society of Architectural Historians*, XL:2, May 1981, pp.168-170(168).

Fletcher's *A History of Architecture*, showing that once criticality is lost, "pluralism" can rapidly slide into an utter "relativism". With its Chinese edition published in 2011, this is a book that once entitled *A History of Architecture on the Comparative Method* and has haunted the mind of two or three generations of modern Chinese architects for nearly a century thanks to its "Tree of Architecture" contained in the book from the 4th to 16th editions. For many modern Chinese architects, the "Tree of Architecture" is a depreciation of Chinese architecture by locating Chinese architecture in the category of "non-historical styles", while delineating a "Western-Centrism" through the legitimacy of "historical styles" — hence the *A Pictorial History of Chinese Architecture* written in 1940 by Liang Ssu-Ch'eng, one of the forerunners of architectural discipline in modern China, as the most solemn yet heroic attempt to counteract the "Tree of Architecture" and the Western-Centrism it represented.

Starting from the seventeenth edition, the succeeding editors of Fletcher's *A History of Architecture* have deleted the distinction between the so-called "historical styles" and the "non-historical styles" in a likely attempt to get rid of "Western-Centrism", if not the Hegelian mainline history in its strict sense. In doing this, unfortunately, it has also relegated itself to nothing other than an encyclopedic chronicle of architecture around the world. This development is unmistakably legible in the 20th edition of the book published in 1996, an edition that the Chinese translation was based on, in which new chapters on architecture of the Middle East, Southeast Asia, Indian Subcontinent, Eastern Europe and Latin America are added up. It is in this edition too that Chinese architecture in the 20th century is imparted with an independent chapter. However, apart from dry historical materials, one can hardly find any critical assessments and discussions of these materials. It seems to its editors and authors, at least with regard to the Chinese chapter, that criticality is no longer necessary, since, as Prof. Luo Xiaowei has discerned, in an affirmative manner though, its point of departure is "relativism"[1].

Having this in mind, one might come to the conclusion that what is important in Curtis' examination of modern architecture is not only that it is emblematic to the shift from "mainline history" and

1 Luo Xiaowei, Preface to the Chinese edition of *Sir Banister Fletcher's A History of Architecture* (trans. Zheng Shiling et al. Beijing: Intellectual Property Publishing House, 2011), pp.i-iv(iv).

"historical determinism" to pluralistic and regional views of architectural history, but also that it reminds us of the necessity to watch out for the danger of relativism as one is taking on that shift. For all its apparently crippled knowledge on modern Chinese architecture, in other words, what makes Curtis' book nevertheless instructive for our own understanding of modern Chinese architecture is its potential to enforce us to think over following questions in a global context of architecture since 1900: for what reasons are such buildings as the Great Hall of the People, the Museum of Chinese History and Revolution, the Building Ministry in Beijing as well as the Chiang Kai-Shek Memorial Hall in Taipei less convincing? Is it about the "skin-deep" touch of tradition alone or there are other issues to think about?

To pose the question by referring to the more recent buildings in China: for what reason is the Chinese Pavilion in Shanghai Expo, a building charged with enormous political and cultural meanings and highly appraised both officially and publically, fails to gain an international recognition that could match its prestige in China? In contrast, what makes Wang Shu's building works recognized and praised worldwide, even though this outcome should in no sense preclude us from critical scrutiny of his works? What are the "regional" or individual in Wang Shu's architecture, and what are the "universal" if the word still makes sense after all? Should we agree with Wang Shu in his assumption of an "indigenous Chinese architecture" that is ultimately to oppose or resist "Western architecture"? Does this "indigenous Chinese architecture" really make sense for the future development of Chinese architecture? And in what way are we to understand Curtis' statement that "it is not unreasonable to posit a 'universalizing' aspect… so long as one strips away the Western bias and progressive assumptions which lurk behind this formulation"[1]?

Seen in light of these and other related issues, and notwithstanding its very limited engagement with modern Chinese architecture, this *magnum opus* by Curtis is certainly a thought-provoking work that should be read widely by the architectural students, professionals and academics who are concerned with the past, present and future of modern Chinese architecture.

---

1 William Curtis, *Modern Architecture since 1900*, p.372.

4

勒·柯布西耶 *Vers une architecture* 译名考[1]

1 最初发表于《新建筑》2014年第2期（总153期）。收录本文集时有修改。

一

勒·柯布西耶有一本著名的论著叫《走向新建筑》，这几乎是中国建筑学尽人皆知的常识。该书法文版原文的标题为 *Vers une architecture*, 从字面理解是“走向一种建筑”的意思。那么，一个原本看似平淡如水的标题何以变成掷地有声的《走向新建筑》？说起来，这得归功于一名叫弗雷德里克·艾切尔斯（Frederick Etchells）的英国艺术家和建筑师，他在 1927 年翻译出版了柯布的这部著作，并以 *Towards a New Architecture* 作为该英文译本的书名。正是艾切尔斯的英文译本使柯布的这部划时代性著作在英语世界并且通过英语在全世界得到广泛传播。一定程度上，中文的《走向新建筑》就是这一影响的结果之一。从 1927 年由汪申伯提出的首个中文译名《向一个建筑去》（显然这个译名带着强烈的法文原文的字面色彩），到吴景祥先生于 1981 年完成的首部中译本采用的书名《走向新建筑》（尽管吴先生的中文版是根据法文版翻译的）[1]，这个译名是如此深入人心，不仅在我们的现代建筑史教材中得到完全认可，[2]

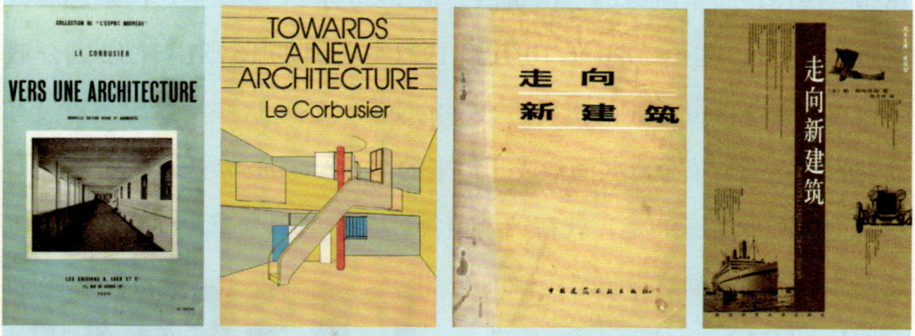

左：勒·柯布西耶 *VERS UNE ARCHITECTURE* 法文版　中左：《走向新建筑》英文版（艾切尔斯译）
中右：《走向新建筑》中文版（吴景祥译）　右：《走向新建筑》中文版（陈志华译）

而且在学术和非学术的语境中得到广泛运用。

从艾切尔斯开始，人们为柯布这部著作的标题加上一个“新”字，无非是要凸显和强调该书不同于以往建筑学论著的可以谓之“新”的内容和特质。这一点在我国学者对该书的论述和评价中也彰显无疑。比如，吴焕加先生在为一直作为全国建筑学主要教材之一的《外国近现代建筑史》撰写的有关文字中这样阐述道（这部分文字后来也出现在吴焕加先生自己的《20 世纪西方建筑史》之中）：

---

1 从汪申伯的《向一个建筑去》开始，该书的中文译名还有《达新建筑之路》（卢毓骏，1932）、《新建筑之路》（卢毓骏，1936）、《到新建筑之路》（《新建筑》渝版第 2 期内附广告，1941）、《走向一个新建筑》（吴景祥，1958），以及 1981 年出版的吴景祥译本《走向新建筑》。《走向新建筑》的译名也被 1991 年出版的陈志华译本采用。

2 见《外国近现代建筑史》（第二版，罗小未主编），北京：中国建筑工业出版社，2004 年版，第 75-76 页。

> 《走向新建筑》是一本宣言式的小册子，里面充满了激奋的甚至狂热的言语，观点也很芜杂，甚至相互矛盾，但中心思想是明确的，就是激烈否定19世纪以来的因循守旧的建筑观点和复古主义、折衷主义的建筑风格，激烈主张创造表现时代的新建筑。[1]

作为具体内容，吴唤加先生简述了该书对现代工业成就的赞美、对工程师美学的推崇、“住房是居住的机器”的论点与大规模住房生产需求的关系、对新的时代需要新的建筑平面的认识、对简单几何形式的热情以及对建筑作为艺术和精神创造的肯定，等等。[2]

诚然，从 *Vers une architecture* 到 *Towards a New Architecture* 再到《走向新建筑》，人们从“新”的角度对其进行阐述和理解不能不说具有相当充分的理由。除了吴唤加先生强调的上述种种“新”特质之外，一个更为直接的理由是该书的大部分文字最初都是在一份名为《新精神》（*L'Esprit nouveau*）的杂志上发表的——事实上，几乎所有学者在论述柯布这部著作时都会谈到它与《新精神》杂志的关系。该杂志创办于1920年10月，其名称源自法国诗人纪尧姆·阿波利奈尔（Guillaume Apollinaire，1880—1918）同年所做的一次讲演。从1920年到1925年，该杂志总共出版了28期，其中1920年10月到1922年5月的16期中有12期登载了柯布与署名Saugnier的奥赞方合写的文章。正是这些文章与一篇没有发表的《建筑与革命》共同形成了后来的 *Vers une architecture*。

然而值得注意的是，就在《新精神》杂志创刊号上，柯布与奥赞方发表的第一篇文章的标题是“Trois rappels à MM les architects”（艾切尔斯和陈志华先生的翻译分别为“Three Reminders to Architects”和《给建筑师先生们的三项备忘》。这个标题之所以值得注意在于作者使用的rappel一词。在法语中，rappel有叫回、召回、想起、提醒的意思，其中一个重要的含义是对过去事物或问题的提醒和召回，更确切地说是对过去曾经有过但现在已经被人们忽视或遗忘的事物的重新提醒和召回。

《新精神》杂志第一期

1 同上。也见吴唤加：《20世纪西方建筑史》，郑州：河南科学技术出版社，1998年版，第104页。

2 同上，第104-105页。

换言之，通过将“体量”“表面”“平面”作为给建筑师先生们的三个 rappels—— 或许应该将其译为“回顾”或者“旧事重提”，柯布强调的与其说是“新”不如说是“旧”。有学者将柯布 1920 年使用的 rappel 与身兼诗人、小说家、剧作家、设计师、编剧、艺术家和导演等多种不同角色的法国人让 · 考克多（Jean Maurice Eugène Clément Cocteau，1889—1963）在 1926 年写的一篇题为“秩序的召回”（Le Rappel à l' Ordre）的短文联系起来。[1] 有趣的是，同样的诉求时隔半个多世纪后再次出现在肯尼斯 · 弗兰姆普敦（Kenneth Frampton）为他后来的《建构文化研究》（*Studies in Tectonic Culture*）所写的一篇纲领性论文的标题之中 ——《召回秩序：建构之境》（“Rappel à l' Ordre: the Case for the Tectonic”），弗兰姆普敦这样表述他的“建构”研究的初衷：

> 之所以选择参与这个议题的讨论是因为我相信，面对建筑表达不断沦为商品的文化趋势，建筑师有必要重新寻求自身的定位。既然大范围直接抵抗难以为继，那么与其迷恋先锋主义的奇思异想，不如采取一种“后卫”(rearguard) 姿态更为恰当。[2]

是的，我们绝不能将 1920 年代风华正茂的柯布与弗兰姆普敦的“批判建筑学”立场（其中包括“批判地域主义”和“建构学”）完全等同起来。但是，在那个现代艺术和现代建筑方兴未艾的时代，柯布却和奥赞方一起在《新精神》杂志撰文给建筑师先生们三个“旧事重提”，这又不能不令人深省。也许，在柯布那里，之所以要“旧事重提”，既不是因为它们“新”，也绝非因为它们“旧”，而是因为它们对于建筑学来说具有某种“本质”的意义。这让人想起勒 · 柯布西耶和奥赞方一起倡导的“纯粹主义”（Purism），一种建立在柏拉图式的“本质”基础之上的现代艺术流派。尽管人们对建筑学“本质“的理解不尽相同，比如弗兰姆普敦就主张，作为建造活动的“建构”是比柯布提出的“表面”“体量”和“平面”更为本质的建筑学内容，[3] 但是这一点至少已经可以说明，在《走向新建筑》的过程中，柯布从来都不是一个价值虚无或者一味崇尚“新”的未来主义者。

关于这部著作的标题，柯布一度曾经想以“建筑还是革命”来命名。他甚至已经为以此为书名的著作设计了封面草图，但后来还是放弃了这

1 Jean-Louis Cohen, “Introduction,” *Toward an Architecture*, translated by John Coodman (Los Angels: Getty Research Institute, 2007), p. 49.

2 Kenneth Frampton, “Rappel à l'Ordre: the Case for the Tectonic,” *Theorizing a New Agenda for Architecture: An Anthology of Architectural theory 1965-1995*, edited by Kate Nesbit (New York: Princeton Architectural Press, 1996), pp.518-528(520).

3 Kenneth Frampton, “Rappel à l'Ordre: the Case for the Tectonic,” p.520.

左：勒 · 柯布西耶《建筑还是革命》封面设计草图
中：更名为 *VERS UNE ARCHITECTURE* 之后的封面设计草图
右：塔基尼亚号海轮

个看似冲击力十足的标题，选择用貌不惊人的 *Vers une architecture* 作为自己第一部理论著作的书名。有趣的是，无论哪一个书名，柯布都在封面上安排了一张阿基塔尼亚号海洋轮船的走道照片。通过这张照片，我们似乎能够领略书名中那个“走向”的意思，然而处在尽端的小方块所代表的就一定是所谓的“新建筑”吗？换言之，将 *Vers une architecture* 译为 *Towards a New Architecture* 或者《走向新建筑》是毫无疑义的吗？

二　事实上，除了给建筑师先生们的三个“旧事重提”，柯布这部著作还有许多方面指向过去的建筑，或者是对过去的重新诠释。比如，针对古老的罗马，该书有“罗马城的教益”一章。该章不仅配上了那张著名的罗马建筑及其几何原型的插图，而且还以这样动人的文字作为开始：

> 人们使用石头、木材、水泥；人们用它们造成了住宅、宫殿；这就是营建。创造性在积极活动着。但，突然间，你打动了我的心，你对我行善，我高兴了，我说：这真美。这就是建筑。艺术就在这里。[1]

当然，该章也不乏对罗马城的批判：

> 罗马太拥挤了。…… 罗马人不懂得怎样使用大理石。…… 罗马的教益

1 勒 · 柯布西耶：《走向新建筑》（陈志华译），西安：陕西师范大学出版社，2004 年版，第 131 页。同上，第 104-105 页。

是为那些明智之士，他们懂得并能欣赏，他们能够抵制，能够克制。罗马是那些知识很少的人的沉沦堕落。把建筑学的学生送到罗马去，这是谋害他们的生命。大罗马奖和美第奇别墅是法国建筑的癌。[1]

常说柯布痛恨以巴黎美院为代表的学院建筑学，这应该没有错。然而一个不争的事实却是，为说明建筑的比例问题，柯布不仅使用了中世纪的巴黎圣母院的案例，而且也毫不犹豫地援引诸多学院派的建筑作品如位于星形广场的凯旋门、凡尔赛的小特里阿农等，当然还有他 1916 年完成的建筑作品施瓦布别墅（Villa Schwob）。建筑史学家约翰·萨默森（John Summerson）曾经将柯布的这种做法视为“古典走向现代”的表征之一。[2] 本文无意将柯布说成一个“古典主义者”，但上述比例分析的案例至少说明，对于写作 *Vers une architecture* 这部著作时期的柯布来说，“现代风格”“新风格”或者“机器美学”还不是那么至高无上，倒是在后来的《作品全集》（*Oeuvre complète*）第一卷（1910 — 1929）之中，已经采用钢筋混凝土框架结构的施瓦布别墅却因不具“现代风格”而未能在这个所谓的“作品全集”中获得一席之地。关于这一点，柯林·罗（Colin Rowe）曾经“一语道破天机”：“该建筑显然与他后来的作品不合拍，倘若与它们收录在一起，将会削弱全集的教化意义。”[3]

学者们注意到，柯布的这部著作还使用了诸多学院派建筑学的术语，

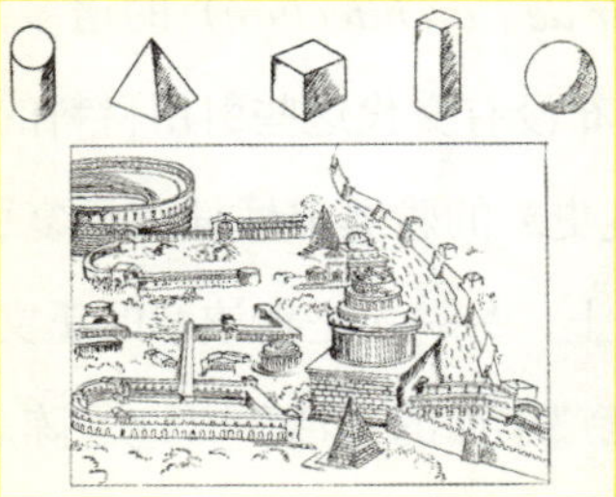

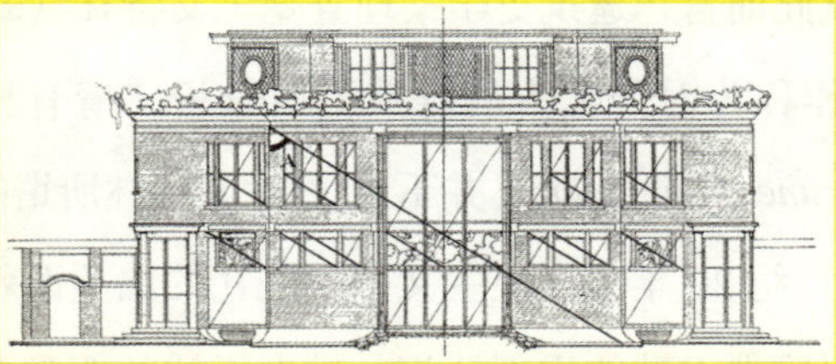

左上：*VERS UNE ARCHITECTURE* 中的罗马建筑与几何原型施瓦布别墅　右上：施瓦布别墅（1916）
左下：*VERS UNE ARCHITECTURE* 中的施瓦布别墅花园立面上的比例分割线
右下：以多米诺体系为起点的《勒·柯布西耶作品全集》第一册

1 勒·柯布西耶：《走向新建筑》（陈志华译），第 132、136、148 页。

2 约翰·萨莫森，《建筑的古典语言》(张欣玮译)，杭州：中国美术学院出版社，1994 年版，第 2 页。同上，第 104-105 页。

3 Colin Rowe, “Mannerism and Modern Architecture,” *The Mathematics and the Ideal Villa and Other Essays* (Cambridge, Massachusetts and London: The MIT Press, 1999), p.30.

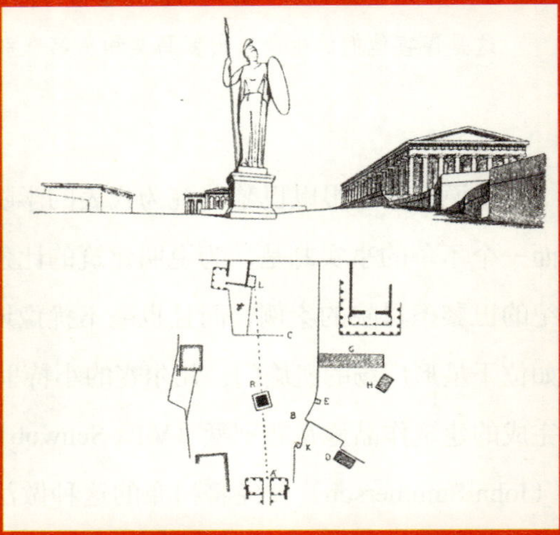

左：舒瓦齐《建筑史》法文版　右：柯布引用的舒瓦齐《建筑史》中的雅典卫城插图

如 ordonnance（法则）、modénature（章法）等，[1] 而柯布那句强调建筑平面设计重要性的著名论断 Le plan est le générateur——“平面是生成器”（艾切尔斯和陈志华先生分别译为 The Plan is the generator 和“平面是生成元”，而吴景祥先生的翻译则是“平面布局是根本”）实际上也可能来自作为学院建筑学理论之代表的朱利安·加代（Julien Guadet）。[2]

与此同时，在作为“给建筑师先生们的三个旧事重提”之三的“平面”一章中，柯布引用了源自 19 世纪法国建筑史学家奥古斯特·舒瓦齐（Auguste Choisy）《建筑史》（*Histoir de l'architecture*）的诸多插图，尽管如同该书的许多其他引用一样，柯布没有交代这些引用材料的来源。更值得一提的是，柯布把舒瓦齐《建筑史》的雅典卫城插图反复引用三次，一次出现在“平面”一章的标题页上，另一次是在该章的正文之中，而第三次则是在论述“建筑”的第二章“平面的花活”（这是陈志华先生的翻译）中出现的。就此而言，建筑史学家理查德·艾特林（Richard Etlin）将柯布的这部著作称为对舒瓦齐的雅典卫城分析“情有独钟”（stands at the heart of *Vers une architecture*）并不为过。[3] 艾特林所谓的“雅典卫城分析”是经过 19 世纪 20 年代浪漫主义革命洗礼的法兰西希腊主义建筑学（Hellenism）对雅典卫城的重新认识。借助于法兰西雅典学院（L'École Française d'Athènes）的成立，一部分学院建筑师的视野成功超越了罗马，转而在雅典卫城看到了一种不同于轴线对称的建筑原则。此

1 John Goodman, “From the Translator,” *Toward an Architecture*, p.xi-xii.

2 Colin Rowe, “La Tourette,” *The Mathematics and the Ideal Villa and Other Essays*, 注释 7。

3 Richard Etlin, “Le Corbusier, Choisy, and French Hellenism,” *The Art Bulletin*, June 1987 Volume LXIX Number 2, pp.264-278(275).

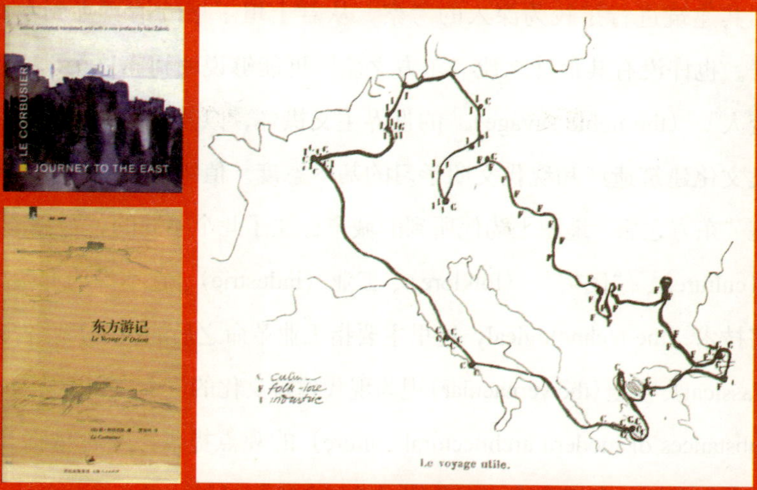

勒·柯布西耶“东方之旅”　左上：《东方之旅》英文版　左下：《东方之旅》中文版
右：“东方之旅”路线图以及对到访城市的三种标注：C- 文化（Culture），F- 民俗（Folklore），I- 工业（Industrie）

前，这个原则曾经在英国人的园林景观中被称为“如画”（picturesque），现在却在“正宗”的古典主义传统中找到例证，其实质是非规则性、无序中的有序以及对场地特殊性的理解和尊重。以此为基础，舒瓦齐提出了雅典卫城作为不规则系列场景设计的理论。一定程度上，正是这一理论启发了柯布“建筑漫步”（promenade architecturale）的思想。[1]

艾特林进一步指出，在 *Vers une architecture* 之中，如同在 19 世纪法兰西希腊主义建筑学中一样，帕提农神庙被视为永恒的美学典范和未来建筑品质的试金石。这固然导致了“建筑：精神的纯粹创造”这一章的主题，但是也与“视而不见的眼睛：汽车”一章有关现代建筑必须像当代的工程建造物一样具有坦率且富有表现力的主张有着千丝万缕的关系。

> 在勒·柯布西耶看来，如同多立克神庙从帕埃斯图姆时期的原始状态发展到帕提农的登峰造极一样，汽车的发展也在不断努力中尽善尽美。[2]

当然，柯布对雅典卫城和帕提农神庙的赞美并非完全只是在步舒瓦齐和 19 世纪希腊主义的后尘，而是充分结合了他自己的现场体验和理解。1911 年，从德国到君士坦丁堡，在穿越欧洲和小亚细亚的旅行期间，柯布与友人奥古斯特·克里普斯坦因（August Klipstein）一起造访雅典卫城。这次旅行后来被柯布称为“东方之旅”（Voyage d’Orient）。[3] 正是通过这

---

1 Etlin, “Le Corbusier, Choisy, and French Hellenism,”，p.273.

2 Ibid，p.275.

3 Le Corbusier, *Journey to the East,* edited, annotated, translated, and with a new preface by Ivan Žaknić (Cambridge, Massachusetts and London: The MIT Press, 2007).

次旅行，柯布将自己的视野拓展到“西方建筑”之外，对“东方”的土耳其建筑进行了较为深入的考察，从君士坦丁堡的清真寺到乡间的民居。也许没有其他什么比“东方之旅”更能够说明柯布这个“高贵的野蛮人”（the noble savage）[1]的世界主义襟怀，以及他那种不拘一格向人类文化建筑遗产和现代文明学习的基本态度。值得注意的是，柯布曾经在“东方之旅”地图上为他所到的城市标注了几个不同的主题：“文化”（culture）、“民俗”（folklore）、工业（industrie）。它不仅为弗兰姆普敦将技术（the technological，这里主要指工业革命之后的技术）、古典（the classical）、乡土（the vernacular）视为现代建筑文化的三个本质要素（essential substances of modern architectural culture）的观点提供了某种依据，[2]而且向我们表明，在柯布建筑生涯的“成型之年”（formative years），“新”和“旧”、“传统”和“现代”就不是彼此孤立甚至对立的。

确实，纵观 *Vers une architecture* 全部六大章的内容，建筑学的新旧内容相互交错且相辅相成。第一章“工程师的美学 · 建筑”论述可以谓之“新”，第二章“给建筑师先生们的三个旧事重提”则是“旧”，第三章“视而不见的眼睛”是“新”，第四章“建筑”则是“旧”，第五章“成批生产的住宅”是“新”，而第六章“建筑或者革命”则希望以“建筑”换取“革命”。可以说，全书章节安排煞费苦心。在这里，所谓“新”只是柯布的诉求之一，而非全部。至少，我们可以这样认为，在柯布看来，只有“新”并不足以完成《走向新建筑》的目标。或者说，除了“新”之外，某种意义上的“旧”仍然是《走向新建筑》不可或缺的。

就此而言，柯林 · 罗将柯布称为“最兼收并蓄而又创造力非凡的折衷主义者”（the most catholic and ingenious of eclectics）[3]的论述可谓精辟。

勒 · 柯布西耶“东方之旅”速写　左：伊斯坦布尔清真寺　右：土耳其民居

1 Adolf Max Vogt, *Le Corbusier, the Noble Savage: Toward an Archaeology of Modernism* (Cambridge, Massachusetts and London: The MIT Press, 1998).

2 Kenneth Frampton, *A Genealogy of Modern Architecture: Comparative Critical Analysis of Built Form* (New York: Lars Müller Publishers, 2015), p.8.

3 Colin Rowe,“ The Mathematics of the Ideal Villa”, p.15.

然而同样重要的是，尽管“兼收并蓄”，柯布却不是一个来者不拒的“收藏家”，这一点在他对“罗马城的教益”的阐述中已经彰显无遗。这就宛如一个来到琳琅满目的市场的淘宝者，需要独具慧眼才能去伪存真，发现有价值的“拾来之物”。那么，柯布的“眼光”意味着什么？这是一个有趣且值得深究的问题，不过也是本文的有限空间无法探讨的问题。

三

让我们回到 *Vers une architecture* 的译名上来。尽管艾切尔斯的英译本 *Towards a New Architecture* 在 1927 年就已经问世，但有趣的是，许多用英语写作的学者在提到柯布这部著作时却常常倾向于使用该书的法文书名，而不是艾切尔斯的英文译名。这类案例不胜枚举，比如，吉迪恩（Sigfried Giedion）最具经典意义的现代建筑史学著作《空间、时间与建筑》（*Space, Time and Architecture*）第一版曾经有一个标题为“作为著者的勒 · 柯布西耶”（Le Corbusier as a writer）的小章节（这一章节在后来的版本中被删除），它在提及柯布这部著作时使用的就是 *Vers une architecture*。[1]《空间、时间与建筑》首版于 1941 年，这已经是艾切尔斯英译本问世后 14 年。我们很难设想，作为现代建筑运动的风云人物和史学家，吉迪恩完全不知道艾切尔斯英译本的存在。那么，如果吉迪恩并非对艾切尔斯英译本一无所知，却仍然选择使用该书的法文书名，这一现象至少说明，尽管吉迪恩曾经为现代运动和现代建筑的立足和发展而不遗余力，但是他对艾切尔斯过于简单化的翻译并不完全认可，个中原因从本文以上论述中也许已经不难理解。

班纳姆《第一机械时代的理论与设计》英文版

关于这一点，建筑史学家雷纳 · 班纳姆（Reyner Banham）的论述无疑值得特别一提。有学者指出，作为一位“未来主义的现代主义”（futurist modernism）的建筑史学家，雷纳 · 班纳姆（Reyner Banham）其实对柯布与学院派藕断丝连的做法一直耿耿于怀，甚至大加鞭挞，[2] 但是他在《第一机械时代的理论与设计》（*Theory and Design in the First Machine Age*）中的论述仍然值得用精辟来形容。在班

1 Sigfried Giedion, *Space, Time and Architecture: the Growth of a New Tradition*（Cambridge: Harvard University Press, first edition, seventh printing, 1947）, p.426.

2 Anthony Vidler, *Histories of the Immediate Present: Inventing Architectural Modernism* (Cambridge, Massachusetts and London: The MIT Press), p.123.

纳姆看来，“勒·柯布西耶的第一本关于建筑的书被证明是20世纪所有建筑论著中最具影响、阅读最广而又是最少得到人们理解的著作之一（one of the most influential, widely read and least understood of all the architectural writings of the twentieth century）。”[1] 他进而指出：

> 如果 Vers une architecture 的原标题只是“走向一种建筑”的意思，而且从书中的内在证据可以看出，它意味着一种一直存在却被遗忘的绝对或者本质的建筑（an absolute or essential architecture, which had always existed and had merely been mislaid）的话，那么英文译本将其译为 Towards a New Architecture 以及德文译本将其译为 Kommende Baukunst（即将到来的建筑）则为整个问题赋予了完全不同的色彩，尽管这一色彩并不能在书中找到充分支持。[2]

班纳姆这里说的德文译本是由德国艺术史学家汉斯·希尔德布兰特（Hans Hildebrandt）于1926年完成的、在时间上早于艾切尔斯的英译本。可以看出，如同对艾切尔斯的英译本书名一样，班纳姆对希尔德布兰特的德译本书名也感到不够满意，认为它们过于强调“新”或“即将到来”的建筑，而柯布的原著却并非完全如此。班纳姆的《第一机械时代的理论与设计》初版于1960年，然而就在此后的1963年，德文译本再版时的书名被改为 *Ausblick auf eine Architektur*（《展望一种建筑》），这基本上可算是法文书名的字面翻译了。这个修改是否考虑了班纳姆的批评意见我们不得而知，但是其用意已经可见一斑。

除了德文版之外，1939年在阿根廷首都布宜诺斯艾利斯出版的西班牙语译本的书名为 *Hacia una arquitectura*，基本上也是法文书名的直接翻译。在日本，宫崎谦三于1929年根据法文版第二版翻译出版了日文译本，书名是《建築芸術へ》（《走向建筑艺术》）。1976年，素有柯布的第二个日本弟子（一般认为，第一个是前川国男）之称的吉阪隆正在1958年的第四版 *Vers une architecture* 的基础上重新翻译，书名为《建築を目指して》（《以建筑为目标》）。这也是柯布这部著作现在日本较为通行的译本。

在所有外文译本中，艾切尔斯的英译本 *Towards a New Architecture* 持续最长，影响也最大。然而，正如建筑史学家威廉·柯蒂斯（William Curtis）曾经指出的，该译名应该用“误译”（mistranslated）来形容，[3] 而在

---

1 Reyner Banham, *Theory and Design in the First Machine Age* (London: Butterworth & Co (Publishers) Ltd., 1960), p.220.

2 Ibid，p.246.

3 William J. R. Curtis, *Modern Architecture since 1900* (London and New York: Phaidon, Third Edition, 1996), p.168. 也见威廉·J·R·柯蒂斯：《20世纪世界建筑史》（本书翻译委员会译），北京：中国建筑工业出版社，2011年版，第168页。

勒·柯布西耶 *VERS UNE ARCHITECTURE* 新版英文版(2007)

此之前，另一位建筑学者斯坦尼劳斯·冯·摩斯（Stanislaus von Moos）则在自己的柯布专著中将 *Towards a New Architecture* 称为“不准确的译名”（the inaccurately translated title）[1]。有理由相信，正是基于对艾切尔斯译本的不满，美国盖蒂研究所（Getty Research Institute）的建筑文献部（Texts & Documents）重新组织翻译，并于 2007 年出版了书名为 *Toward an Architecture* 的新版英文译本。译者约翰·古德曼（John Goodman）在“译者的话”（From the Translator）中指出：

> 艾切尔捕捉到了这部著作的断奏基调，但为了读起来顺口而使它的措辞变得均质化，模糊了该书强烈的意志性以及魔咒般的节奏。柯布的这部著作其实要比艾切尔的英译本难得多——修辞上更为思辨且充满暗示。[2]

相较于语气的差异，术语的不准确或许是一个更为严重的问题。古德曼特别指出，艾切尔斯将柯布这部著作中数次出现的颇具学院派色彩的术语 ordonnance（法则）译为 arrangement（安排、布置），这无疑值得商榷，因为后者“不具备那种规范性的力量，因而也传递了一种错误信息，它常常指向一种随意的道德方式，一种无需遭受惩罚的处世规则”。[3]

同样，法国建筑史学家让 - 路易·柯恩（Jean-Louis Cohen）也在他为新版英译本所写的长篇“引言”（Introduction）中指出：

> 在对待 *Vers une architecture* 时，他（指艾切尔斯，引者注）更像一位宣传家而非批评家。他相当不准确地将该书的标题译为 *Towards a New Architecture*，还毫不顾忌地用自己的 12 页引言取代了 1924 年版

1 Stanislaus von Moos, *Le Corbusier, Elelments of a Synthesis* (Cambridge, Mass. and London: The MIT Press, 1979), p.46.

2 John Goodman: "From the Translator", p. xi-xii(xi).

3 Ibid. p.xii.

本上的柯布前言，并配备了相当平庸的英国建筑和结构以及一个手术室的照片。他承认自己的翻译“有些句子笨拙可笑，保留了一些高卢主义”。[1]

柯恩还指出，艾切尔斯英译本“最令人尴尬的不准确用词”（the most emblematic of the many embarrassing textual inaccuracies）是“将‘给建筑师先生们的三个旧事重提’之一的‘体积’（volume）变成‘体块’（mass），这招致了赖特对该书的批评”。[2] 也许，在赖特看来，对于建筑而言，“体量”和“体块”不可混淆，前者是具有内部空间的，而后者则趋向于一个实心的块状物，这无疑与赖特的“有机建筑”理念相去甚远。很显然，如果赖特的批评是成立的，那么他批评的其实是艾切尔斯的英译本，而让柯布背负由此产生的指责无疑是莫大的冤枉。换言之，赖特对柯布著作的批评更像是一个误会，而这个误会恰恰是由艾切尔斯的英译本引起的。不幸的是，这个“误会”似乎也在陈志华先生的译本中成为事实。在那里，与吴景祥先生译本中的“体量”不同，“给建筑师先生们的三个备忘”之一正是“体块”，尽管在陈译中文版封面之后出现的是盖有清华大学藏书章的法文版扉页！

四

那么，如果源自艾切尔斯 *Towards a New Architecture* 的《走向新建筑》不是一个令人满意的译名的话，应该如何翻译更好？在笔者看来，有以下几种可能：1）《走向一种建筑》，这在字面上最为贴近原文，也是西班牙文译本和新版英译本（以及某种意义上 1963 年德文版）采取的策略。但是这是怎样一种，还是可以理解为任意一种，则是一个有待思考的历史理论问题。2）《走向本质的建筑》，这多少受到班纳姆观点的影响，认为 *Vers une architecture* 思考的是“一种一直存在但却没有得到充分理解的绝对或者本质的建筑”。在笔者看来，这一做法的危险程度与《走向新建筑》的翻译大抵相等，即译者的诠释过多，尊重原文不够。3）《走向一种建筑观》，这比较接近 1963 年德文版中那个 Ausblick 的意思，强调的是一种观点，一种理念。当然，这样翻译必然会凸显一个“观”字，不如新版英文译名或者以上第一种翻译方法在字面上更贴近原文。4）《走向建筑》，它省略了与原文不定冠词相

1 Jean-Louis Cohen, “Introduction”, 1-78(49).

2 Ibid.

对应的“一种”。这样做并非不恰当，因为西方语言中的不定冠词并不一定都需要在中文中翻译出来。比如，意大利建筑史学家布鲁诺·赛维（Bruno Zevi）曾经有一部名为 *Towards an Organic Architecture* 的著作，中文一般将其译为《走向有机建筑》，而不是《走向一种有机建筑》。同样，挪威建筑理论家和历史学家克里斯蒂安·诺伯格-舒尔茨（Christian Norberg-Schulz）的 *Genius Loci: Towards a Phenomenology of Architecture* 就不一定译为《场所精神——走向一种建筑现象学》，尽管事实上它意味着一种有别于其他建筑现象学观念的“建筑现象学”[1]。一定意义上，《走向建筑》更类似吉阪隆正的《建築を目指して》（《以建筑为目标》），也与台湾 1997 年版本的《迈向建筑》[2] 所表达的思路比较吻合。作为一个普通名词，这里的“建筑”有别于其他属类的事物或概念，如文学、音乐或者革命，这一点不仅在该书最后一章的标题“建筑或者革命”中得到反映，而且柯布也确实一度想用《建筑还是革命》作为全书的书名，尽管我们并不能因此得出柯布这部著作的全部用意只是阐述在“建筑”和“革命”之间二选一的结论。将 *Vers une architecture* 译为《走向建筑》的另个好处是，尽管没有用“新”或者“本质”进行修饰，但是它在中文语境中仍然具有某种“宣言”的力度，而这种力度正是翻译柯布这部著作时无论如何都不能忽视的。

结语

勒·柯布西耶的建筑之路不可能没有“新”，他在 *Vers une architecture* 问世之后三年的 1926 年提出著名的“新建筑五点”（Les 5 Points d’une architecture nouvelle）更突显了对“新”的强调——有理由相信，后者可能就是艾切尔斯将 *Vers une architecture* 译为 *Towards a New Architecture* 的诱因之一。确实，这个“新建筑五点”不仅构成了他 1920 年代纯粹主义小住宅建筑的实质内容，而且也反映在他所谓的《作品全集》其实并非“全集”，而是作为“新建筑五点”之基础的多米诺体系（*Maison Dom-ino*）被提出之后的“作品全集”这一事实之中。然而 *Vers une architecture* 是否应该译为 *Towards a New Architecture* 或者《走向新建筑》的疑问其实并不在于柯布建筑是否包含“新”，而在于所谓“新”是否就是柯布建筑的全部，更在于《走向新建筑》的译名极易引发的误解和认识上的偏差。

---

1 参见本文集《意义的探究——克里斯蒂安·诺伯格-舒尔茨建筑理论评述》和《都市田园中的建筑悬浮——评斯蒂文·霍尔的深圳万科中心》。

2 Le Corbusier：《迈向建筑》（施植明译），台北·田园城市出版社，1997 年版．

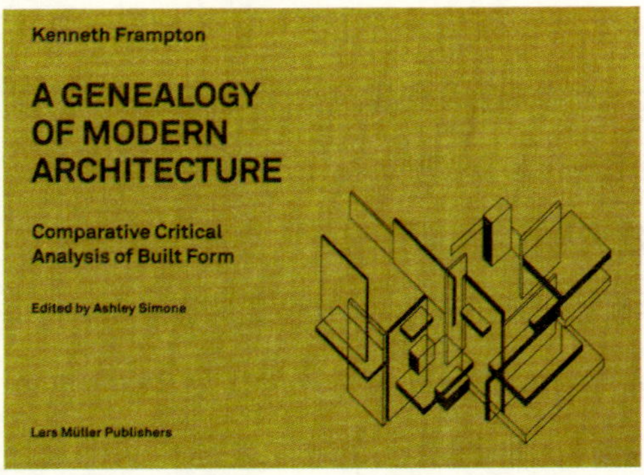

肯尼斯·弗兰姆普敦《现代建筑谱系——建筑形式的批判性比较分析》英文版

事实上，正如弗兰姆普敦在《现代建筑谱系》（*A Genealogy of Modern Architecture*）中指出的，“新建筑五点”提出四年之后的1930年，柯布设计的智利的艾拉索利斯住宅（Errazuris House）似乎“在一夜之间抛弃了”（abandoned overnight）由这五点所定义的“新建筑”，转而尝试一种“乡土现代主义”（vernacular modernism）。弗兰姆普敦指出，在此后的1930年代，尽管柯布完成了诸如巴黎的救世军大楼（Armée du Salut）和日内瓦的克拉泰住宅（Maison Clarté）这类机械主义建筑作品，他同时也在探索一条将乡土建造模式与工业化生产元素相结合的建筑道路。1935年建成的周末住宅（Maison Week-End）就是这一探索的结果之一。[1] 如果弗兰姆普顿将技术（the technological，这里主要指工业革命之后的技术）、古典（the classical）、乡土（the vernacular）视为现代建筑文化的三个本质要素（essential substances of modern architectural culture）的观点是准确的，那么现代建筑历史上也许没有什么比柯布的建筑生涯更能说明这一点。

然而，弗兰姆普敦所谓的“现代建筑文化”确实既不同于19世纪的学院派建筑——无论柯布从学院派那里有多少继承，这一点在 *Vers une architecture* 中也是显而易见的，也不同于20世纪80年代开始兴起的后现代主义，尽管这一点看似与柯布没有任何直接关系。站在21世纪的今

勒·柯布西耶位于智利的艾拉索利斯住宅（1930）

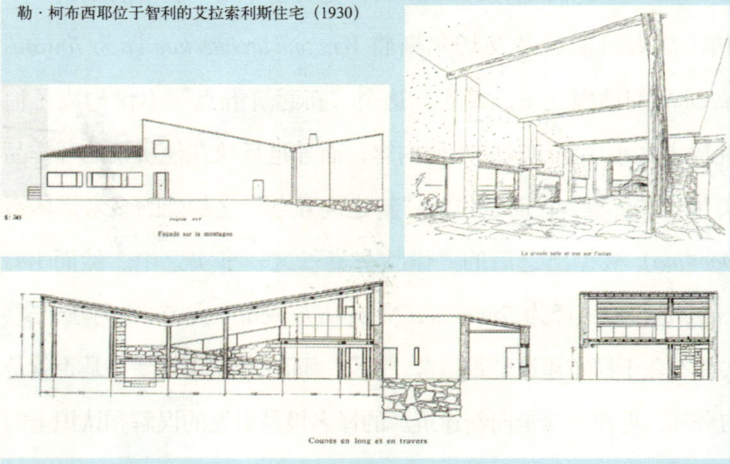

1 Kenneth Frampton: *A Genealogy of Modern Architecture*, p.12.

天看，“现代建筑文化”已经成为一种传统，一种在 *Vers une architecture* 中已经初现端倪，或者用弗兰姆普敦的话来说，一种在 1920 年代中期到 1980 年代早期之间发展而成的现代建筑中得到体现的“新传统”（tradition of the new）[1]。它既是技术、古典和乡土的融合 —— 就此而言，艾切尔斯的 *Towards a New Architecture* 以及由此引发的《走向新建筑》的中文翻译就过于简单和片面，甚至引起误解，也意味着一种与 20 世纪之前的建筑以及 20 世纪 80 年代之后出现的后现代主义迥然不同的特殊的现代主义的建筑主张。笔者以为，在后面这层意义上来说，《走向一种建筑》也许反倒是上述数种选择中看似最貌不惊人但实则最贴近柯布原意的译名。在这里，与其说它画蛇添足地保留了通常在中文翻译中可以省略的“一种”，倒不如说它这样做的目的是为读者留下某种悬念。至于这个“一种建筑”在柯布那里到底意味着什么，人们自己去理解和感悟吧。就像柯布在该书封面设计上使用的阿基塔尼亚号海洋轮船照片走道尽端的小方块所隐喻的那样，它可能是“新”的，也可能是“旧”的；可能是“精神的”，也可能是“物质的”；可能是“普世的”，也可能是“地域的”；可能是“兼收并蓄的”“混合的”“折衷的”，也可能是“绝对的”“本质的”，或者用吉迪恩晚年的话来说，是一种“永恒的现在”（the eternal present）[2]。这些都是柯布建筑的特质，而这大概也是柯布及其作品的魅力经久不衰的原因。然而需要再次强调的是，“融合”并非这一魅力的关键所在，因为如同后现代主义曾经声称“融合”一样，人们也太容易将现代与传统的“融合”挂在嘴边。真正的魅力在于在 *Vers une architecture* 中已经呼之欲出的实现这一融合的勒·柯布西耶方式。

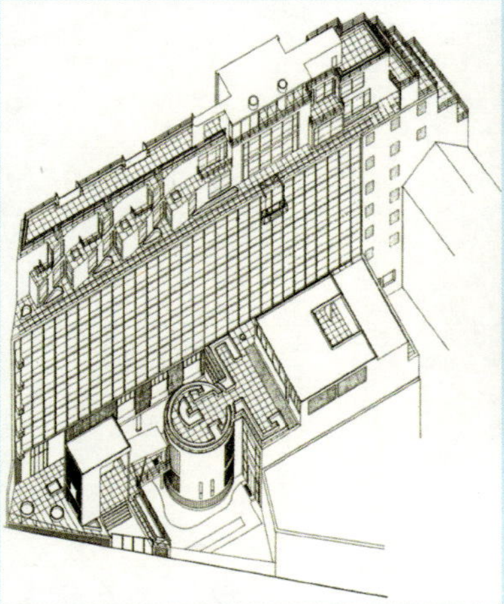

勒·柯布西耶　左：救世军总部大楼　右：周末住宅（1935）

1 Frampton: *A Genealogy of Modern Architecture*，pp.6-7.

2 Sigfried Giedion, *The Eternal Present: A Contribution on Constance and Change*（New York: Pantheon Books, 1962）.

5

空间、建造、表皮

——论赫佐格与德默隆早期的建筑艺术 [1]

1 最初以"空间，构造，表皮与极少主义——关于赫佐格和德默隆建筑艺术的几点思考"发表于《建筑师》1998年第5期（总84期）。收入本文集有修改。

引子

密斯有一句名言："少就是多"，即以英文形式广为流传的"Less is More"。尽管从史学的角度并没有人提供过足够的证据可以说明，为什么以德语为母语的密斯会有一句英语的格言（密斯常说的德语口头禅是 beinahe Nichts，直译"几乎什么也没有"），人们还是普遍将一度以密斯传人自居的菲利普·约翰逊 (Philip Johnson) 在他那本于 1947 年首次出版的关于密斯的小册子中所说的这句密斯名言作为事实来接受了。在这本小册子里，约翰逊对"少就是多"以及密斯的建筑主张不乏赞颂之词，这在当时是很具代表性的。

然而，1960 年代之后，尤其在"后现代主义"风头正劲的日子里，"少就是多"几乎成了密斯的一大罪状。罗伯特·文丘里（Robert Venturi）在他那本带有宣言性质的《建筑的复杂性与矛盾性》（*Complexity and Contradiction in Architecture*）一书中就特别将批判的矛头指向密斯及其"少就是多"。文丘里写道：

> 建筑师再也不能被清教徒式的正统现代主义建筑的说教吓唬住了。我喜欢基本要素混杂而不要"纯粹"，妥协而不要"干净"，扭曲而不要"直率"，含糊而不要"条理清晰"，既反常又无太多个人色彩，既乏味又"有趣"，宁要平凡而不要过分的设计味儿，容纳而不是排斥，丰富而不是简单，既要旧的也要新的，宁可不一致和不肯定也不要直接和明确的。[1]

在文丘里看来，与他所倡导的"杂乱而有活力的包容性统一"和"既此又彼的二元论"相反，少是"轻而易举的统一和排他性"的表征。他的结论是"多并不是少""少使人厌烦"。

正如人们不应将 20 世纪尤其是二战之后粗制滥造、千篇一律、单调乏味的建筑归罪于"少就是多"的思想一样，笔者也无意在这里将后现代主义时代那些繁琐堆砌、缺少节制的建筑仅仅视为文丘里所倡导的"矛

罗伯特·文丘里《建筑的复杂性与矛盾性》英文版

1 罗伯特·文丘里：《建筑的复杂性与矛盾性》（周卜颐译），北京：中国水利水电出版社／知识产权出版社，2006 年版，第 16 页。

左：《少就是多》展览目录　　右：唐纳德·贾德《无题》

盾性与复杂性”的直接后果。值得强调的反倒是，在后现代主义思潮几乎风靡全球的时候，却涌现了一批以不同国家和地区为基地孜孜不倦地探索现代主义之后在不同区域条件下重寻建筑之本质和纯净形式的建筑师。墨西哥的路易斯·巴拉干 (Louis Barragán), 英国的大卫·奇普菲尔德 (David Chipperfield), 葡萄牙的爱德华多·索托·德·莫拉 (Eduardo Souto de Moura)，以及本文论及的被一些评论家称为“北方瑞士”或“德语瑞士”建筑师之一的雅克·赫佐格 (Jacques Herzog) 和皮埃尔·德默隆 (Pierre de Meuron) 当属此列。

毋庸讳言，我们这里涉及的不是什么运动或流派一类的东西。如果说在这些建筑师之间有什么共同之处的话，那就是对建筑的挚爱以及通过简练的形式、纯净的空间和精巧的结构来追求和表现这种挚爱的实践。在这样的意义上，这些建筑师体现的也是建筑中的一个深远的或许可以称为极少主义 (minimalism) 的传统。

1996 年为配合在巴塞罗那举行的第 14 届世界建筑师代表大会，西班牙加泰罗尼亚建筑学院举办了一个题为“少就是多”的展览，将这一传统直接追溯到密斯那里。[1] 然而正如纽约现代艺术博物馆（MoMA）设计部主任和策展人特伦斯·瑞利 (Terence Riley) 曾经指出的，抛开对后世具有深远影响的古希腊柏拉图思想中极少主义的因素不谈，西方建筑中的极少主义至少可以上溯到 12 世纪欧洲西多 (Cistercian) 宗教改革运动中那些与当时占统治地位的充满装饰的哥特教堂形成鲜明对照的修道院建筑。这种由于宗教热情而导致的极少主义后来在美国颤震派教徒

1 *Less is More: Minimalism in Architecture and Other Arts*, catalogue directed by Vittorio E. Savi & Josep Maria Montane (Barcelona: Col.legi d'Arquitectes de Catalunya ACTAR, 1996).

(Shaker) 的文化和建筑中也有极致的表达。[1]

不可否认，自从工业革命给技术、材料等带来深刻变化之后，也许没有哪一位早期现代主义建筑师像密斯那样在其作品中充分表现了现代极少主义的精神，他的一些作品也已成为 20 世纪极少主义的经典之作。然而对于赫佐格和德默隆来说，来自 20 世纪 60 年代美国极少主义艺术，尤其是极少主义雕塑的影响肯定是最大的，尽管人们可以在他们的作品里看到以上其他极少主义传统以及诸如概念艺术 (Conceptual Art) 和波普艺术 (Pop Art) 等现代艺术的影响。

早在 70 年代赫佐格就对唐纳德 · 贾德（Donald Judd）、罗伯特 · 莫里斯（Robert Morris）、理查德 · 塞拉（Richrd Serra）的艺术产生了浓厚的兴趣。此后他和德默隆与瑞士艺术家莱米 · 曹格 (Rémy Zaugg) 富有成果的合作也已成为理解其建筑作品的不可缺少的因素。当然，作为 20 世纪晚期的极少主义建构诗人，赫尔佐格和德默隆的作品还须以建筑自己的内容才能得以充分说明。笔者以为这样的内容至少包括三个方面，即空间、建造和表皮。

空间

建筑与空间有着本质的关系，或者说建筑是空间的创造，这从人类开始进行营造活动之日起大概就是无可争辩的事实。然而，据建筑史学家彼得 · 柯林斯 (Peter Collins) 考证，“空间”作为一个明确的概念直到 18 世纪才真正出现在西方的建筑论著中。这是一个颇为奇怪的现象，因为即使抛开别的不说，意大利文艺复兴时期对透视法的研究没有明确的空间概念是完全不可设想的。本文无意追究这一现象的原因，[2] 更值得注意的是使“空间”一词真正在建筑论著中成为常用词的主要是 18 世纪在德国兴起的艺术史学家们。

这一点之所以值得注意是因为“空间”一旦为史学家们所用，它立刻成为一个历时性的概念，即历史上每一个时期的建筑空间都是在与其他时期的空间相比较才得以充分阐述的。从温克尔曼（Johan Joachim Winckelmann）到沃尔夫林 (Heinrich Wölfflin) 均是如此。到 20 世纪这一艺术史的传统直接为现代建筑的史学家们所承用。吉迪恩 (Sigfried Giedion) 是这样，而赛维 (Bruno Zevi) 则有过之而无不及，意欲以空间为主题来完成一部建筑的历史。赛维这样写道：

1 Terence Riley, “Introduction,” *Waro Kishi* (Barcelona: Editorial Gustavo Gili, S.A.,1995), pp.6-12(6).

2 可以说，原因是复杂多样的，关于柯林斯从语言学角度提供的解释，见《现代建筑设计思想的演变》（英若聪译），北京：中国建筑工业出版社。

> 直到目前，还没有一本令人满意的建筑史著作问世，原因是我们还不习惯从“空间”的角度来思考问题。[1]

那么，赛维是如何论述这样的历史的呢？在赛维看来，古希腊的神庙建筑固然伟大，它却有忽视内部空间的缺陷。空间，准确地说内部空间真正成为西方建筑的主角始于古罗马时期，而这一时期的空间的基本特征就在于它的静态性。赛维写道：

> 不论圆形还是方形这两种空间，其共同规律是对称性，与相邻空间的关系都是绝对各自独立的，厚重的分隔墙越发加强了这种独立性，以超人的宏伟尺度构成双轴线的壮观效果，基本上是不同于观者而在效果上会有任何变化的、沉静独立的存在。[2]

在后来西方建筑的发展中，古罗马时期的这种静态空间经过蛮族入侵时期空间与节奏的间断处理，拜占庭时期节奏急促并向外扩展的空间，罗马风（Romanesque）富有格律的空间，哥特时期对空间向度和连续性的追求，文艺复兴空间透视的意识以及巴洛克空间的动感和渗透性逐步为现代建筑的有机空间所代替。那么，从赛维的观点看，什么是现代建筑的有机空间？赛维指出：

> 现代建筑空间是以开放的平面为基础的。……有钢铁和混凝土的新结构技术，可以将支承部分缩减为纤细的骨架，这就提供了实现灵活或开放平面理论的现实条件。……使用了大窗户，现在是整片墙都做成玻璃的，这就使室内空间和室外空间达到完全连续的效果。[3]

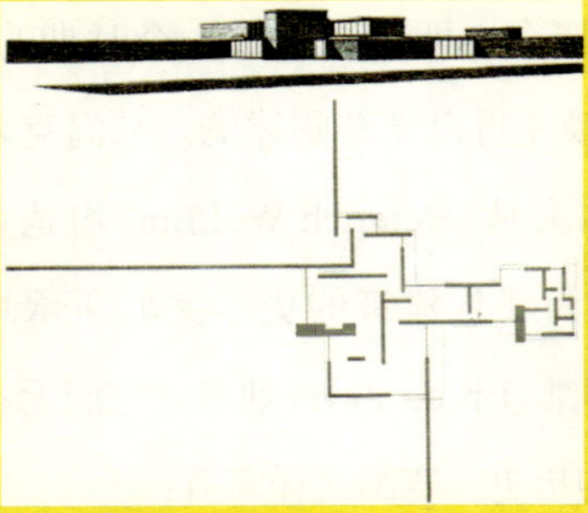

左：赛维《建筑空间论》中文版（张似赞译）　中：密斯　砖宅　右：凡·杜斯堡　时空中的建造

1 布鲁诺·赛维：《建筑空间论》（张似赞译），北京：中国建筑工业出版社，1985 年版，第 9 页。
2 同上，第 45 页。
3 同上，第 82 页。

在西方建筑的历史上，也许没有哪一个时代像现代主义时期一样通过史学的方法达到如此强烈的自我意识，而空间无疑是这一意识的重要组成部分。当然，我们不能把现代建筑空间意识的形成完全视为史学家工作的结果，毋宁说史学家们的工作在一定意义上只是对现代建筑实践的总结和阐述。事实上，早在赛维完成他的《建筑空间论》前数十年，密斯就在他设计的砖宅方案中探讨了流通空间的新概念，而这一方案本身又深受“风格派”（De Stijl）特别是凡·杜斯堡（Theo van Doesburg）的“时间－空间的建造”（The Construction of Space-Time）的影响。

如果说砖宅还只是一个方案的话，那么 1929 年的巴塞罗那博览会德国馆则以一个建成作品一举震惊了当时的建筑世界。它的墙体自由布置，空间处处流动，平屋顶出檐深远，轻盈潇洒，整个建筑无论在平面的二维和立体的三维上均向四方舒展延伸，室内室外融为一体。另一方面，如果说巴塞罗那德国馆的围合部分主要还是以实体性的大理石墙面构成的话，那么密斯后来在发展“通用空间”(total space)或“一统空间”(universal space) 等概念时，他的建筑的围合部分则多为完全透明的大玻璃，其极致的表现除了伊利诺理工学院建筑系馆和柏林新国家美术馆之外，大概就要算范斯沃斯住宅（Farnsworth House）。在这些建筑中，不仅内部实现了“通用”或“一统”空间，而且事实上内部空间近乎消失，达到赛维所谓的室内外空间完全连续的效果。

之所以在这里不厌其烦地重述现代建筑空间的历史和理论，乃是因为我们这一代建筑师多少是以接受现代建筑的原理开始学习建筑的。但是，当我们把空间的流来流去以及室内室外的相互交融看成是建筑艺术的核心，并且在过去的近二十年里又耳闻目睹所谓“大 P 小 m”的 Postmodernists（如文丘里，查尔斯·摩尔，格雷夫斯等）用历史符号和

密斯·凡·德·罗　范斯沃斯住宅

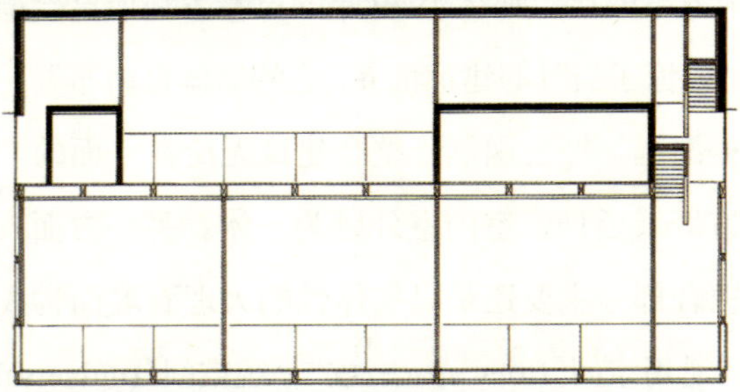

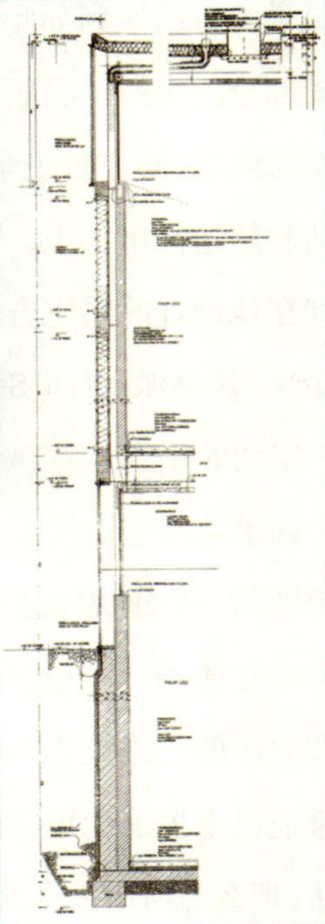

右上：戈兹美术馆　左下：剖面　　右：剖面大样

装饰将空间渲染得唯恐不够丰富，或者“小 p 大 M”的 post Modernists（如屈米、哈迪德、艾森曼等）追求空间的变形、扭曲、倾斜之后，竟在赫佐格和德默隆的建筑中惊奇地看到，一个既不流通也不有机，既无装饰也无符号，既不变形也不扭曲，四四方方围合起来的空间原来也会有如此震撼人心的力量。

或许，在赫佐格和德默隆众多的建筑作品里，没有比坐落在德国慕尼黑的戈兹美术收藏馆 (Goetz Collection) 更能充分显示这一点。这是一个只有三千余平方米的小型私人美术馆。它的基地位于慕尼黑的一片别墅区里，与其主人的住房相邻，周围环境绿荫葱葱，幽雅宁静。在这样的环境里赫佐格和德默隆构造了一个十分小巧精致的建筑物。从外表看它是一个地地道道的盒子式建筑。它的立面在垂直方向分为三段，上下为玻璃，中间为木质胶合板。除了个别细部外，这种垂直方向的三段式划分几乎毫无变化地为建筑的四个立面所采用。

与建筑外部集装箱似的形象一致，它的内部空间也是简单而又规则的。展室的分隔并不是结构上的承重墙，但分布规规矩矩，近乎刻板，一点也没有现代建筑自由平面的精神。各展室的空间既不互相渗透流通也不与室外交融，而是实实在在的独立存在。更值得一提的是，与过去数十年中为数不少的、设计者使尽浑身解数意欲创造建筑艺术但结果却往往喧宾夺主将展品沦为建筑的点缀的博物馆建筑相比，戈兹美术收藏馆的展室可谓朴素无华。它的墙面上方为高窗，使自然光进入展室。墙面洁白淡雅，木地板平整光洁，天花除平整的灯槽之外别无他物。在这里，艺术品获得了充分展示自身美的机会和场所。

凡·德·兰　法尔斯修道院

显然，这是一个带有极少主义意味的空间。但是，如果说就雕塑而言极少主义的成功之处就在于“它发现了这样一种积极空间，即当观赏者在空间里运动时可以意识到自身感知大大改变物体作为艺术品的意义”的话，[1] 那么戈兹美术收藏馆的室内则是向赛维所谓的罗马静态空间的回归。当然，它并没有赛维在罗马静态空间里所看到的“超人的宏伟尺度”和“双轴线的壮观效果”。就此而言，戈兹美术收藏馆或许更准确地说是令人想起西多宗教改革运动中的极少主义空间。这种空间曾经被荷兰修道士和建筑师汉斯·凡·德·兰（Hans van der Lann）在他为法尔斯修道院设计的建筑中得到了精彩的再现[1]。不过，与凡·德·兰的修道院不一样的是，戈兹美术收藏馆的空间既是最基本的，其建造又是 20 世纪钢筋混凝土结构、轻型木结构以及幕墙玻璃的混合结果。当然，如同大多数德国 / 瑞士建筑一样，它也充分体现了当代材料和工艺技术的精确之美，其最终结果的特征无疑是 20 世纪而非 12 世纪的。

建造

赫佐格和德默隆同生于 1950 年，并于同一年（1975 年）从瑞士苏黎世高工（ETH）建筑系毕业。这是一所由伟大的德国 19 世纪建筑师哥特弗里德·森佩尔 (Gottfried Semper) 于 1855 年创建的学校，在世界众多建筑院校中素以重视建造 (konstruieren) 教学而闻名。1978 年赫佐格和德默隆共创建筑事务所，此后就一直以瑞士城市巴塞尔（Basel）为基地从事建筑活动。说起巴塞尔，人们不免会想起哲学家尼采。就在这里，年方 25 岁的尼采成为古典语言的教授。尼采对古希腊文化的推崇众所周知。在他看来，古希腊人具有一种率直自然、朴实无华的前基督教精神。这种精神将身体和建筑的完美性置于崇高的

1 Glòria Picazo,“Minimalist Space and Time,” *Less is More: Minimalism in Architecture and Other Arts*，p.125.

地位。身体的完美性体现在古希腊人崇尚思想的自由和健魄的身躯的理念之中，而建筑的完美性则与精致的建造不可分。要想求证赫佐格和德默隆的思想里有多少尼采的直接影响也许并非易事，但有一点可以肯定，建立在精致构造基础之上的建造不仅是苏黎世高工建筑系的优良传统，而且也是瑞士文化的重要组成部分，这无疑在它诸如钟表业悠久的历史中充分体现出来。可以说，正是基于一个远自古希腊和工艺制造业的传统，进入 20 世纪 90 年代的欧洲建筑思想界对简单性的讨论同时也与精致性、秩序、耐久性等主题紧密结合在一起。

问题的提出始于 1993 年德国《明镜周刊》(*Der Spiegel*) 第 51 期刊载的一篇题为《来自日常的挑战 —— 倡导一种新的建筑传统》（“Die Provokation des Alltäglichen. Für eine neue Konvention des Bauens”）的文章。该文的作者是当时任德国法兰克福德意志建筑博物馆馆长的维多里奥 · 马尼亚戈 · 兰普尼亚尼 (Vittorio Magnago Lampuniani)。在这篇文章中，兰氏从社会的角度分析了当时德国建筑界一方面为后现代主义的历史游戏所左右，另一方面又为解构主义所诱惑的状况。据兰氏看，这一状况不能仅仅以国际建筑潮流来解释。在德国它有更深刻的社会原因。纳粹时代对简洁结实的古典建筑的崇拜使人们对古典传统中真正本质的东西讳莫如深，以致那些当今在建筑中使用石材或木材的同时运用简单几何体块和追求精致细部的建筑师竟有“反动”和“法西斯”之嫌。无独有偶，艾森曼也曾将赫佐格和德默隆这一类的所谓“北方瑞士”或“德语瑞士”建筑师称为“瑞士法西斯”[1]。但是，兰氏认为，经过仔细推敲的和坚固持久的建筑从社会和经济的角度来说却正是我们所需要的。

同年，在德意志建筑博物馆 1993 年年鉴里，兰氏又将类似的思想进一步整理成以“新的简单性”(Die neue Einfachkeit) 为题的建筑六点，其中第一点就是简洁及其在资源、技术和美学上的理由。兰氏强调指出，简单性并不是贫乏，而是提炼和浓缩后的丰富。此后，兰氏在 1995 年的著作《持久的现代性》(*Die Modernität des Dauerhaften*) 中进一步阐述了与之相关的其他主题，如“精致”（Genaugkeit）、“缓慢”（Langsamkeit）、“平静”（Ruhe）、“秩序”（Ordnung）、

兰普尼亚尼《持久的现代性》德文版

1 “Minimalism and Ornament.Herzog & de Meuron in conversation with Nikolaus Kuhnert and Angelika Schnell,” *ARCH+* Dezember 1995, pp.115-118(115).

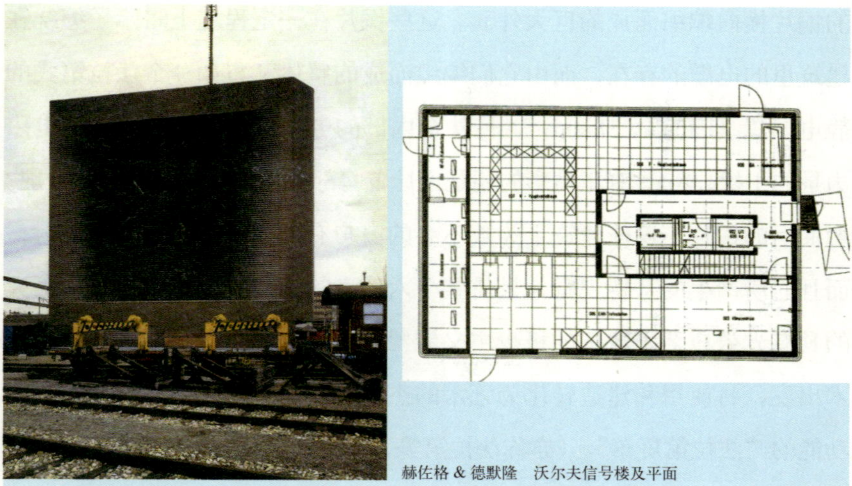

赫佐格 & 德默隆　沃尔夫信号楼及平面

“重力”（Schwere）、“艰难的轻”（die schwierige Leichtigkeit）、“坚固”（Festigkeit）、“建造”（Konstruktion）、“饰面”（Verkleidung）、“功用”（Nützlichkeit）、甚至“设计与生态”（Entwurf und Ökologie）、“节俭与奢华”（Sparsamkeit und Luxus），等等。兰氏反复强调的是建筑作品作为工艺技巧的建构文化，以及这一文化所具有的坚固和耐久等品质在“可持续发展”时代的内在意义。[1]

1992—1996 年间，兰氏还担任意大利设计杂志 *Domus* 的主编。这期间，*Domus* 不仅改变了装帧设计，也在内容上展现出强烈个性。用兰氏自己的话来说，“它要成为一种新建筑的代言人”，而所谓“新”就在于一代青年建筑师在新的条件下对建筑作为建造工艺所持的强烈兴趣和热忱。与此相一致，*Domus* 的方向“就是展示设计和建造均优秀的作品及其过程”。[2]

鉴于以上几个的背景，我们不难理解赫佐格和德默隆的作品会成为 *Domus* 杂志的“常用曲目”。从一开始，除戈兹美术收藏馆外，他们最引人注目的作品大概就要数坐落在巴塞尔铁路调车场的沃尔夫信号楼 (Signal Box auf dem Wolf)。顾名思义，这是一座为监视和调度进出调车场的货车而设立的建筑。从它的顶部人们可以将整个调车场一览无余。然而在一个电脑化的信息时代，真正的调度却是通过电脑的屏幕来进行的。

总的来说，该建筑的功能并不复杂。复杂的倒是建筑中千丝万缕的电缆和它们之间的相互关系。不过这一点在建筑形象上并没有表现出来，至少没有直接表现出来。相反，人们看到的是一个整洁的、由 20 厘米宽

---

1 Vittorio Magnago Lampugnani, *Die Modernität des Dauerhaften, Essays zu Stadt, Architektur und Design* (Berlin: Verlag Klaus Wagenbach, 1995).

2 *Domus*，1994-1.

的铜片横向编织而成的巨大外套。这些铜片在一定程度上暗示了缠绕在建筑里的电缆的存在，而由它们编织而成的整体又宛如一个法拉第式的静电屏罩（Faraday Shield），使建筑内部的电器设备免受外部众多电磁力量的干扰。然而，真正精彩的还是设计者将部分铜片做微妙弯曲的处理。弯曲后的铜片不仅使建筑上为数不多的窗户有机会获得自然光的照射，而且它所产生的奇特视觉效果使整个六层高的信号楼俨然成为一个巨大的具有光效应艺术特征的极少主义雕塑。可以说，这是一个从功能和技术出发，将使用和建造转化为艺术的杰作。事实上，它那具有静电屏蔽功能的“法拉第屏罩”（亦称法拉第笼，Faraday Cage）的设计就曾是赫佐格和德默隆母校苏黎世高工建筑系低年级建造教学的经典范例之一。

如果说沃尔夫信号楼的精彩之处在于它的外表皮及其构造处理的话，那么戈兹美术收藏馆的最不同凡响之处则是它的结构体系。如前所言，它的立面在垂直方向采取了三段式划分。乍看上去，这使人产生建筑内部空间为三层的错觉，而这一错觉又为实质处于与室外地面同一标高并且因为透明玻璃而从外部可以直接看到的夹层设备间和图书室而加剧。事实上，该建筑内部主要空间只有二层：底层为半地下层，与地面接触的玻璃仅为它的高窗部分，而真正的二层则从立面的三分之一高度开始。半地下层的采用使建筑的立面比例更为优雅，夹层的采用使辅助空间与展示空间不同的高度要求得到满足，由此而来的层数的错觉则为该建筑简单的体形增加了某种诙谐性。然而值得注意的是，这种诙谐性并非简单的表面文章。很大程度上，它来自建筑自身独特的结构体系。

从外部看，建筑顶部高窗有节奏的竖向划分和磨砂玻璃背后隐约可见的柱子令人想起通常的框架结构。但是，二层的柱子并没有出现在地面层。真正的结构体系是：与室外地面处于同一标高的辅助空间（入口接待、图书室及设备间）是两个钢筋混凝土筒体，它们不对称地跨置在嵌入地下的钢筋混凝土箱体之上，再用这两个筒体承载二层展厅的轻型木结构。换言之，二层展览空间的结构是完全独立的体系，它整体地落在底层部分的两个混凝土筒体上，而底层则是完全没有柱子的结构，地面层“连续”的磨砂玻璃幕墙以及上部建筑的整体“悬浮”由此产生。

因此，貌似简单的建筑体形却有着并不简单的结构体系，少中有多的艺术可见一斑。当然，如果以肯尼思·弗兰姆普敦(Kenneth Frampton)《建构文化研究》(*Studies in Tectonic Culture*)的建构观点看，戈兹美术收藏馆结构和形式的关系也许是一个值得争议的问题。与他的前人爱德华·塞克勒（Eduard Sekler）一样，弗兰姆普敦在“建构的”（the tectonic）和“非建构的”（the atectonic）之间做出了区分。“建构的”固然不能以结构和构造为唯一内容，因为同样的结构或构造可以有不同的表现形式，但无论如何结构和构造要得到表现，甚至是直接的表现才是符合建构文化的。与此相反，“非建构的”则是“那些将建筑中支撑和被支撑之间具有表现力的相互作用关系在视觉上忽视或使其含糊不清的手法”。[1]

据此，弗兰姆普敦将巴塞罗那德国馆十字形钢柱与梁的交接被密斯为追求平滑的空间流通以及“几乎什么也没有”的极少主义效果而增加的遮挡平整天花板的做法称为“非建构的”或“对建构的压抑”（the suppression of the tectonic）[2]。也许出于同样的理由，他对赫佐格和德默隆的作品持保留态度，甚至予以谴责。[3] 弗兰姆普敦的批评不是没有道理的，因为除了结构问题外，赫佐格和德默隆在创作中对“表皮”的极端重视似乎已经脱离了弗兰姆普敦的“建构文化”所能接收的范围，而在试验与自己的艺术趣味相符合的建构之路。

表皮

关于自己的建筑作品，赫佐格和德默隆曾经说过一句令人回味的话：“当我们想要将人们的注意力引向建筑表皮的时候，我们总是选择简单的建筑形体。”[4] 换言之，就赫佐格和德默隆的建筑作品几乎全是简单形体这一点来说，表皮（surface 或 skin）在他们的创作中的重要性是不言而喻的。在戈兹美术收藏馆那里，除底层的某些部分因为是清水玻璃而可以从外部看进去外，立面的其他部分不是实墙就是半透明的磨砂玻璃。人们如果不将注意力停留在它的表皮就等于不在看这幢建筑。而沃尔夫信号楼的情况就更不用说了：其“法拉第屏罩”作为一个精彩的“构造”则完全是裹在主楼外面的一张皮，它与建筑内部的结构似乎并没有太多的本质关系。

或许，我们有必要回顾一下极少主义艺术与表皮的关系。关于这个问题，艺术评论家罗莎琳·克劳斯（Rosalind Krauss）曾经在一次有关密斯的研讨会上这样说道：极少主义的实质并不在于那种一成不变的几何形体中心。相反，“正由于它一反艺术作品只有通过显示某个隐性中心或者作为其表现，才能获得自身意义的观念，所以它把意义赋予物体的

1 肯尼斯·弗兰姆普敦：建构文化研究——论19世纪和20世纪建筑中的建造诗学（王骏阳译），北京：中国建筑工业出版社，2007年版，第20页。

2 Kenneth Frampton, *Studies in Tectonic Culture: The Poetics of Construction in Nineteenth and Twentieth Century Architecture* (Cambridge, Massachuesetts, London, England, The MIT Press, 1995) p.179．中文版此处译为“对建构的损害”，见《建构文化研究》，第182页。

3 这里的“谴责”（condemned）是赫佐格和德默隆在采访中的表述，见“Minimalism and Ornament.Herzog & de Meuron in conversation with Nikolaus Kuhnert and Angelika Schnell”, p.117.

4 Ibid, p.117.

表面并对反射性材料和挖掘自然光的效果感兴趣”[1]。对于建筑，此话怎讲？换言之，对于建筑来说，表皮的意义何在？

西方建筑中的表皮处理有着悠久的历史。从古埃及的壁刻到古希腊的浮雕，从拜占庭的马赛克画到哥特教堂的彩窗，不一而足。到文艺复兴时期，阿尔伯蒂尝试将建筑的主题（如柱子、拱券等）以及其他母题以大理石镶嵌的方式进行处理，从而将建筑的物体性转化为以表皮来组织的视觉秩序。他设计的鲁切拉伊府邸（Pallazo Rucellai）和新圣母教堂（S. Maria Novella）的主立面都是这方面的经典案例。文艺复兴以后到学院派，严格意义上的阿尔伯蒂式的二维立面虽不多见，但建筑的表皮处理如内部的天花藻井、外部立面的壁柱线脚等仍是建筑师设计工作的重要内容。到 19 世纪末、20 世纪初维也纳建筑师约瑟夫 · 霍夫曼（Josef Hoffmann）和奥托 · 瓦格纳（Otto Wagner）等又对建筑表面做古典母题以外的处理，不过很快随着现代建筑的兴起，体量和空间的重要性大大超过了表皮。这一点在倡导现代主义建筑的勒 · 柯布西耶那里也许最为明显。我们知道，勒 · 柯布西耶曾经提出了著名的“新建筑五点”。人们不太提及的是他在《走向一种建筑》（*Vers une architecture*，通常译为《走向新建筑》[2]）一书中还回顾了“体量”“表皮”和“平面”这三个基本要素，并把它们称为“给建筑师先生们的三项备忘”（Trois rappels à MM les architects）。在勒 · 柯布西耶看来，体量作为纯粹几何体块的创造是建筑艺术的基础，平面是建立三维体量的手段或生成元。但是在柯布那里，如果不对表皮进行处理的话，我们似乎并不能真正达到建筑。表皮的处理可以强化或破坏体量，建筑师的任务属于前者。勒 · 柯布西耶写道：“体块被表面包裹，…… 建筑师的任务是使包裹在体块之外的

左：鲁拉切伊府邸　中：新圣母教堂　右：奥托 · 瓦格纳 奥地利储蓄银行

1 Rosaline Krauss, “The Grid, the /Cloud/, and the Detail” in *The Presence of Mies*, ed.Detlef Mertins (New York: Princeton Architectural Press, 1994), p.134.

2 关于这个问题的讨论，参见本文集《勒 · 柯布西埃 Vers une architecture 译名考》。

表面生动起来，防止它们成为寄生虫，遮没了体块并为它们的利益而把体块吃掉。”[1]就此而言，虽然乍看起来柯布给予表皮以足够的重视，因为表皮成了建筑的三个备忘之一，但是实际上表皮在柯氏那里并没有独立的地位，它充其量只是为体量服务的。

勒·柯布西耶是阳光灿烂的地中海文化的宠儿。他的建筑充分体现了体积的美，光和影的美。问题在于，正如赫佐格曾经写到的，“勒·柯布西埃……把建筑看成是‘阳光下科学性的、正确无误的体量游戏’。但是，如果建筑根本不是游戏，特别不是科学性的和正确无误的游戏，如果太阳并不像在南欧那样时时光芒四射，而是经常被云层遮盖，光线呈弥散状，那情况又会怎样呢？”[2]

在一定程度，赫佐格和德默隆的建筑可以说是对以上问题的思考和解答。首先，如果说探索光的效果作为取得物体多重感知的基本手段对于勒·柯布西耶以及赫佐格和德默隆都适用的话，那么与柯布强烈光影下的塑性建筑不一样，赫佐格和德默隆的建筑体现的往往是更为细微的光效果。我们前面已经提到戈兹美术收藏馆是极为简单的长方体，它的外表完全没有凹凸，因而也就没有强烈的光影和体积效果，有的只是平坦的表面和不同材质的展现。后面这一点很重要。就戈兹美术收藏馆而言，胶合板、清水玻璃、磨砂玻璃、铝合金等不同材料使乍看上去体形简单的建筑上有了细微而丰富的差异，而弥散性的光线正是将不同材料的物质性和随时间所发生的变化视觉化并赋予这种差异以意义的前提条件。

在戈兹美术收藏馆表皮众多的材料中，尤其值得一提的是磨砂玻璃的使用。毫无疑问，磨砂玻璃在该建筑中有其特定的功能作用。一方面，它能够为展室提供充足而又柔和的自然光，另一方面，用两层玻璃将二层支撑结构夹在中间有一定的隔热作用，是当今于福斯特式的高技派提高能耗效率之外的直接式（即从技术上说不采用机械装置）节能方法之一。但是，磨砂玻璃的建筑意义和双层玻璃的采用不能仅用这两点来解释。作为一种材料，磨砂玻璃的透明度介于实墙（是古典建筑的主要元素）和透明玻璃（为现代主义所钟爱）。它的半透明性既使室内室外空间有明确的界限，又避免了传统建筑过于沉闷的围合。另外，正如特伦斯·瑞利指出过的，对于观者来说，它可以将处于另一边的建筑内容若隐若现，含而不露。宛如古罗马美艳的王后波蓓娅（Sabina Poppaea）的面纱，它展现的是一种含蓄而暧昧的、双关的乃至色欲 (erotic) 的美。[3]与之不同，

1 勒·柯布西埃：《走向新建筑》（陈志华译），天津·天津科学技术出版社，1991 年版，第 32-33 页。

2 Jacques Herzog, “The Hidden Geometry of Nature” in *Quaderns*, 1989 No.pp.181-182.

3 Terence Riley, “Introduction”, *Light Construction* (New York: The Museum of Modern Art, 1995). 也见 *The Light Construction Reader*, ed. by Todd Gannon and Jeffery Kipnis (New York: The Monacelli Press, 2002), pp.23-41(26).

赫佐格 & 德默隆　利可乐车间与仓储中心

它的立面上的中段实墙在木构框架之间固定的双层桦木板（双层桦木板之间填充保温材料）的外层上抹灰，同时刻意显露墙体的木构框架。

在赫佐格和德默隆的作品中，戈兹美术收藏馆和沃尔夫信号楼代表了建筑表皮处理的不同策略。前者或是通过材料的纯粹物质性赋予表皮以特征（磨砂玻璃），或是通过表面处理（抹灰）改变人们对材料的感知，而后者则将构造和表皮合二为一。而在利可乐公司（Ricola）的第二个生产车间和仓储中心，我们则可以发现第三种处理表皮的方法，即在严格二维的意义上将某种图案主题通过丝网印刷法在立面重复使用来改变表皮的性质。该建筑位于法国城市米卢斯 (Mulhouse) 南郊一片树木茂盛的地区。按利可乐公司的要求，该建筑既是厂房又兼做库房。对此，建筑师提供的解决方法是简单的大空间加灵活的分隔。从外部看，整个建筑宛如一个超尺度的打开盖子的纸箱子。它的两个悬挑出来的大檐口似乎既在向周围自然环境张开双臂，又给入口部分的装卸工作带来遮阳避雨的功能。建筑的两个侧立面为素混凝土墙体。该部分没有女儿墙，屋顶的雨水沿着墙面流下，日久自然形成图案。与此不同，建筑的前后两个主要立面则为轻巧得多的低造价工业材料聚碳酸酯板。它的透明性能够为车间和仓储中心的工作区域提供经过一定过滤的自然光。设计者对这两个立面做了重点处理，其方法是在它的上面用丝网印刷法将摄影艺术家卡尔·勃罗斯费尔 (Karl Blossfeldt) 的一个树叶作品作为母题进行重复。重复的母题从墙面一直延续到整个挑檐的内侧。

在严格二维的意义上处理建筑的表皮，这使我们回到阿尔伯蒂的传统。在 20 世纪的今天后现代主义建筑师们也采用类似的手法。曾将建筑定义为“具有装饰的屋子”（decorated shed）的文丘里就是其中一位。但是，正如西班牙建筑师亚历杭德罗·萨艾拉（Alejandro Zaera）指出的，赫佐格和德默隆是独特的：如果说从阿尔伯蒂到文丘里表皮的主题是在既定的语言和再现范式 (linguistic-representative paradigm) 内进行操作的话，那么赫佐格和德默隆选择通常与建筑无关的母题（在利可乐新厂房那里为树叶）以及将这一母题重复到极致，是通过将具象的元素淹没在纹理中，从而超越原有的语言学法则（linguistic code）以及母题的能指与所指的关系。毋宁说赫佐格和德默隆的策略是艺术家安迪·沃霍（Andy Warhol）式的，即用重复、韵律和夸大尺度来改变意象固有的意义和再现功能。[1]

继现代建筑的几何抽象性之后，西方建筑理论界不乏重返以历史主义为基础的“具象建筑”的呼声。[2] 无疑，赫佐格和德默隆没有加入这股潮流中去。他们的建筑始终保持着现代建筑的抽象性。但是，他们又离白色派的几何游戏远去。他们的建筑介于抽象和具象之间。他们在探索自己的路。他们是独特的。

结语

让我们回到文丘里的“建筑的矛盾性和复杂性”。据文丘里说，建筑的这一特性或者他对建筑的这一特性的偏爱是“以包括艺术固有的经验在内的丰富不定的现代经验为基础的”。[3] 但是，正如俄国至上主义者（Suprematist）马莱维奇 (Kasimir Malevich) 早就指出过的，艺术品并不必然需要采取华丽多彩的形式才具有丰富的含义。事实上，20 世纪最为优秀的极少主义艺术作品，从埃里克·萨蒂 (Eric Satie)、约翰·凯奇 (John Cage)、 菲利浦·格拉斯 (Philip Glass) 的音乐到萨缪尔·贝克特 (Samuel Beckett) 的戏剧，从雷蒙·卡佛 (Raymond Carver) 和欧内斯特·海明威 (Ernest Hemingway) 的小说到弗兰克·斯特拉（Frank Stella）和巴内特·纽曼（Barnet Newman）的绘画所提供的经验都是如此。

毋庸讳言，我们处在一个日趋复杂的世界。进入 1990 年代以来，建

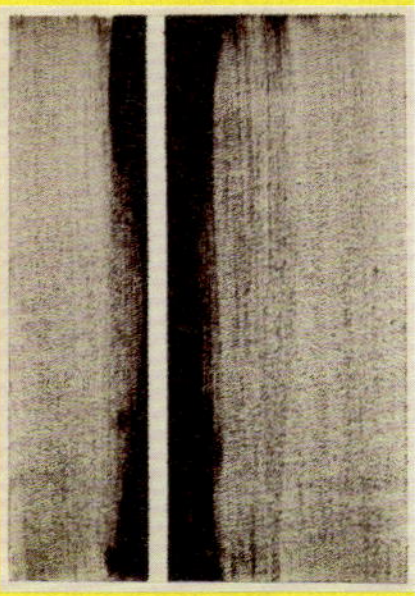

巴内特·纽曼绘画

1 Alejandro Zaera, “Between the Face and the Landscape”, *Herzog & de Meuron 1983-1993, El Croquis*, 1993，pp.24-36.

2 参见本文集《意义的探究——克里斯蒂安·诺伯格 - 舒尔茨建筑理论评述》的相关内容。

3 文丘里：《建筑的复杂性与矛盾性》，第 16 页。

筑界对复杂性的关注已不再是文丘里式的历史主义所能涵盖，然而强调建筑对复杂性的表现却同样是以现代世界之复杂性的理念为其合法性基础的。当年以倡导建筑中的后现代主义而著名的查尔斯·詹克斯（Charles Jencks）近年又出新书并为该书定了一个响亮的副标题：《一种理论：复杂科学是如何改变建筑和文化的》（"A Polemic: How Complexity Science is Changing Architecture and Culture"）。[1] 笔者无意否认复杂性问题对于建筑的重要性，然而对于文丘里提出的"多并不是少""少令人厌烦"的命题而言，通过赫佐格和德默隆建筑作品的分析，我们至少可以相当肯定地说：少中可以有多，少并不一定令人厌烦。

1 Charles Jencks, The Architecture of the Jumping Universe (London: Academy Editions, 1995).

6

都会田园中的建筑悬浮

——评斯蒂文·霍尔的深圳万科中心[1]

1 最初发表于《时代建筑》2010 年第 4 期（总 114 期），收录本文集时有修改。

建筑学

有那么一个世界，在那里，一切都是确切的、几何的、智识的、理性的和合乎逻辑推论的，那是柏拉图的“理念”，是笛卡尔的“我思”；还有那么一个世界，在那里，确切、规整、理性和逻辑的光辉黯然失色，取而代之的是模糊、暧昧、经验、情感、身体、欲望乃至性爱，那是诗的领域，是艺术的本源，在一定意义上也是“现象学”，尤其是梅洛-庞蒂“知觉现象学”的世界。

意大利文艺复兴，一个旨在让古典再生的时代，一代人文主义者阿尔伯蒂以一部《建筑十书》（*De re aedificatoria libri decem*）雄心勃勃地试图重构事实上只是以残篇传世的维特鲁威的《建筑十书》，为西方建筑学确立自己的学科边界奠定基础。“在阿尔伯蒂看来，维特鲁威最重要的成就，无疑在于他的关于比例方面的理论贡献”，[1] 而阿尔伯蒂自己提出的“美”的原则似乎比维特鲁威的比例理论更能表明建筑学科的理论核心在上述两个世界中所处的位置。在这里，“几何将元素统一起来，形成自足的整体，它服从阿尔伯蒂‘不能多也不能少，否则就会破坏整体’的原则。”[2]

与此同时，或者更准确地说是在阿尔伯蒂的《论建筑》问世后近半个世纪的1499年，意大利文艺复兴出现了一部称得上另类的建筑学著作——《寻爱绮梦》（*Hypnerotomachia Poliphili*），或者根据后来出现的英文译本的标题，又称《波里菲利之梦——梦中爱的纷争》（*Hypnerotomachia Poliphili*: *The Strife of Love in a Dream*）。这是一部秉承古希腊对爱欲（Eros）的诗性想象以及中世纪和文艺复兴早期性爱文学传统的游记，它记述了主人公波里菲利（其名字也可解作“许多事情的爱人”的意思）在寻求他的爱人波莉娅（Polia，其名字也可解作“许多事情”的意思）的探险过程中所经历的如梦似幻的销魂世界和性爱体验，而作为一部建筑学理

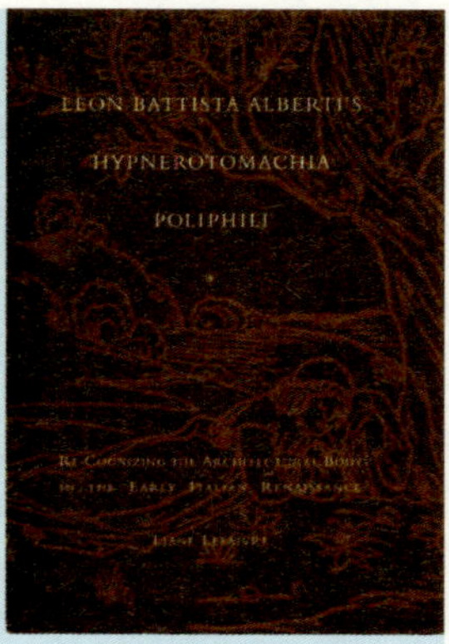

左：阿尔伯蒂《建筑十书》英文版　右：丽安娜·勒菲弗尔《列昂·巴蒂斯塔·阿尔伯蒂的寻爱绮梦》英文版，书中论证《寻爱绮梦》的作者就是阿尔伯蒂！

1 彼得·默里：《文艺复兴建筑》（王贵祥译），北京：中国建筑工业出版社，1999年版，第25-26页。

2 Christian Norberg-Schulz, *Meaning in Western Architecture* (London: Studio Vista, 1980), p.113.

论著作，它的理论基础正是在这样的语境中得到呈现的。

关于这部建筑学奇书的作者，史学家们众说纷纭，莫衷一是。有说无名氏的，也有说是来自威尼斯的方济各会修士弗朗西斯科·科罗纳（Francesco Colonna）的。20 世纪 90 年代，更有学者经过研究考证后语出惊人：作为对《建筑十书》的补充，该书的作者不是别人，正是奠定现代西方建筑学理论基础的阿尔伯蒂！[1] 在此，我们没有时间和空间对这一观点的可靠性进行论证，但是倘若史实果真如此，倒也从一个侧面证实了如下一点：即便在阿尔伯蒂那里（更不用说在更为广泛的文艺复兴人文主义那里），建筑学的理论核心也从未只是一个智识的、建立在比例关系之上的“形式化”的抽象美学价值问题，而是在同样程度上也植根于欲望、梦幻、想象、知觉、身体乃至性爱体验之中。换言之，在阿尔伯蒂的建筑学学科领域中，智识与身体、视觉与体验、客观与主体、几何与诗性、学科规范与个体想象并非截然相对，而是相辅相成的。正如建筑史学家和理论家佩雷兹 - 戈麦兹（Alberto Pérez-Gómez）曾经指出的，只是到了 18 世纪末和 19 世纪初，文艺复兴建筑思想中的多元性才被工具理性的一元性所代替，后者在以迪朗（Jean-Louis Durand）的《简明建筑学教程》（*Précis des Leçons d'Architecture*）为代表的建筑学理论中被系统地确立起来，成为 20 世纪功能主义和形式操作主义建筑学的基础。[2] 在这一过程中，不仅文艺复兴比例理论的形而上含义丧失殆尽，而且“波里菲利之梦”的建筑学意义也逐渐为人们所忘却。

然而，如果建筑学仍然与意义和诗性相关，而这又不能以重返形而上学和神学世界为前提，那么梅洛 - 庞蒂知觉现象学中那个“拥有的身体”（owned body）就成为这一意义和诗性的终极源泉。或许，这就是佩雷兹 - 戈麦兹重返波里菲利之梦中领悟到的“基本的现象学教益”，而由此而来的“基于爱的建造”（built upon love）的建筑学理论则可谓是戈氏为改变当代建筑学开出的一剂“解药”。[3] 值得指出的是，这个所谓“解药”的意义既不仅仅在于它呼吁用身体的经验和诗性的情感来弥补和纠正工具理性、实用主义和拜物主义的不足，也不能因为它视“欲望作为意义

1 见 Liane Lefaivre, *Leon Battista Alberti's Hypnerotomachia Poliphili: Re-cognizing the Architectural Body in the Early Italian Renaissance* (Cambridge, Massachusetts and London: The MIT Press,1997), Chapter Five: The Real Poliphilo.

2 见 Alberto Pérez-Gómez, *Architecture and the Crisis of Modern Science* (Cambridge, Massachusetts and London: The MIT Press, 1983).

3 见 Alberto Pérez-Gómez, *Polyphilo or the Dark Forest Revisited: An Erotic Epiphany of Architecture*（Cambridge, Massachusetts and London：The MIT Press, 1992）。“《波里菲利之梦》基本的现象学教益”之陈述可特别见该书“绪论”第 XXIV 页。也见 Alberto Pérez-Gómez, *Built upon Love: Architectural Longing after Ethics and Aesthetics*（Cambridge, Massachusetts and London： The MIT Press, 2004）。在这里，“解药”（antidote）是芬兰建筑师和建筑理论家约翰尼·帕拉斯玛（Juhani Pallasmaa）对《基于爱的建造》的称谓，见该书封底评论。关于这个问题的讨论，也见本文集《理论何为？——关于建筑理论教学的思考》一文。

之源”（desire as the origin of meaning）[1]而将它等同于放纵和随心所欲的建筑学。爱，是一个伦理学概念。“基于爱的建造”既是为维特鲁威意义上的原本与爱神维纳斯（Venus）有着内在关联的“美观”（Venustas）一词正名，也在一个物欲横流、对大自然进行肆无忌惮的掠夺而导致日益严重的环境危机的时代具备了某种更为深层的伦理学意义。诚如该书的编者所言：“佩雷兹-戈麦兹揭示爱与建筑的关系就是旨在寻找介于诗性与伦理之间——介于建筑师设计美丽世界的愿望与建筑学为社会创造更好场所的迫切任务之间的结合点。”[2]

悬浮

从1994年的*A+U*特集《感知的问题——建筑现象学》（*Questions of Perception: Phenomenology of Architecture*）开始，斯蒂文·霍尔的名字就在公众意识中与建筑现象学联系在一起了，尽管在一定意义上，霍尔的“现象学转向”在之前的《锚固》（*Anchoring*）时期已经形成。[3]在这一过程中，正是佩雷兹-戈麦兹和芬兰人约翰尼·帕拉斯玛（Juhani Pallasmaa）与霍尔结成了“建筑现象学”的“锵锵三人行”，共同倡导更为感性的、具有多重知觉特征（触觉、听觉，乃至嗅觉，而非仅仅视觉）的建筑观，对抗工具理性主义的建筑学和空洞的形式主义游戏。三人的相同之处大概已经毋须多说，更有意思的反倒是霍尔与其他二人的差异。这一差异既不在于霍尔是一位有大量建筑作品问世的建筑师，而其他两人则主要以理论著作为人所知，也不在于霍尔的理论思维常常与自己的建筑作品交织在一起，因而缺少其他两人著作中那种深邃的学术性和历史感。在“建筑现象学”的道路上，霍尔的独特性在于他对理念的重视和挖掘。

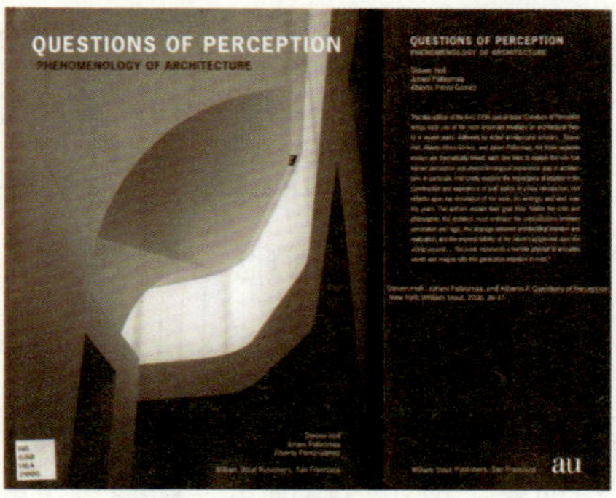

A+U 专辑《感知的问题——建筑现象学》

理念（idea），在柏拉图的意义上是现象的对立。对于柏拉图来说，

1 Alberto Pérez-Gómez, *Polyphilo or the Dark Forest Revisited*, p.xvi.Alberto Pérez-Gómez, Polyphilo or the Dark Forest Revisited, p.xvi.

2 Alberto Pérez-Gómez, *Built upon Love*，封底。

3 显然，尽管在《锚固》中“建筑现象学”还没有明确提出，但是某种意义上的“现象学意识”已经呼之欲出了。沈克宁先生指出，在《锚固》中，霍尔的现象学思想多少还是存在现象学式的，而在1991-1992年左右阅读了梅洛-庞蒂的《知觉现象学》之后，霍尔的现象学思想发生了变化，开始走向知觉的建筑现象学。见沈克宁：《建筑现象学》，北京：中国建筑工业出版社，2008年版，第3页。

理念是“原型”和绝对的真，而现象则属于“原型”的个别显现，因而不可信。但是在霍尔那里，柏拉图式的理念与现象的对立已经不复存在。取而代之的是，正如人们在霍尔那里反复可以看到的，理念通常在概念中得到发展，它既是揭示不同现象特性和经验特性的有效工具，又可借助艺术、科学、哲学的特殊术语和名词突破日常经验和既有建筑观念的限制和束缚，成为激发新的设计思维的催化剂。因此，在霍尔建筑思想的发展过程中，人们不仅可以发现“锚固”“交织”“重”“轻”“光”“声音”“城市边缘”等在日常经验面产生的概念，还可以看到“视差”（parallax）、“奇异吸引子”（strange attractor）、“弹性视域”（elastic horizon）、“染色空间”（chromatic space）、“孟结海绵”（Menger sponge）、“黏滞性”（viscosity）、“多孔性”（porosity）、“染色体交叉”（chiasma）、“影子的速度”（speed of shadow）、“物质的化学”（chemistry of matter）等源自天文学、气象学、光学、动力学、生物学、物理学、化学等学科的术语和名词，更不要说诸如“绵延”（duration）、“纠缠的经历”（enmeshed experience）等颇具柏格森哲学意味的概念。可以说，正是从这些术语、名词、概念中衍生出来的理念与充满感性而又神秘色彩的水彩画形成了霍尔建筑思想（说“现象学思想”也无妨）的两个基石，它们将智识与体验、理性与情感、经验与抽象、现实与超现实、精确与想象融为一体，再通过杰出的设计才华和独特的个人语言，铸就了众多富有创造性和想象力的建筑作品。

进入 21 世纪以后，霍尔开始涉足中国项目。先是 2002 年的南宁柳沙半岛概念性规划竞赛，接下来是 2003 年分别在北京和南京进行并进入实施的当代 MOMA 和佛手湖建筑艺术博物馆（当然还有更多的项目曾经或者正在进行之中）。[1] 在这些项目中，早先的理念有些依然清晰可见，另一些则若有若无，甚至了无踪影。这倒也没有什么奇怪。我们既不能期望一位建筑师永远停留在一种或几种理念上面，也不能要求他在一个设计中

霍尔　左：《锚固》英文版　中：《交织》英文版　右：《视差》英文版

1 见《世界建筑》2007 年第 1 期。

囊括自己的所有理念。与此同时，一些新的理念似乎在霍尔的中国项目中发展出来，其中“悬浮”无疑是最具诗意也最具建筑表现性的一个。

在一个被万有引力支配的世界，“悬浮”永远与想象、自由、超越、欲望、诗意、梦想紧密联系在一起。自古以来，西方有“天使”，东方有“飞天”，它们都是人类想象和欲望的产物，都是人类追求自由与超越的诗性表达。但是，如果“天使”的形象强调的是翅膀的力量，那么尽管中国古代也有“羽人”之说，而且这种“羽人”也在敦煌石窟中屡见不鲜，但是中国的“飞天”中出现更多的却是“不长翅膀，不生羽毛，没有圆光，借助云彩而不依靠云彩，主要凭借一条长长的舞带而凌空翱翔的‘仙人’”。[1] 换言之，中国式“飞天”无需翅膀，却可以腾云驾雾般飘飘欲仙，自由来往。这是一种更为优美的“悬浮”，也是一种更加人化、但又更具诗意的想象。因此，当人类凭借科技的发展已经完全实现了物理飞行之后，当“天使”随着“上帝”的隐退不再动人的时候，身体的悬浮——不是西方传统绘画中那种顶着显灵光圈的圣人的悬浮，而是现实人的悬浮——反倒成为现代艺术中最具诗意的表现主题之一，它在夏加尔（Marc Chagall）、马格利特（René Magritte）、拉蒂克（Jacques-Henri Lartique）、达利（Salvador Dalí）的作品中都曾出现过。贝拉·夏加尔这样描述蕴含在马克·夏加尔的绘画《恋人》中的诗意：“……你腾空而起，伸展着四肢，向天花板冉冉漂浮起来，你转过头，也把我的头转过来，……咱们俩就联袂在点缀好的房间里飞升，我们终于飞起来了。在窗前我们真想穿越而过。窗外的白云和蓝天在召唤着我们。”[2] 在这里，即使没有波提切利《维纳斯的诞生》中的翅膀，“恋人”们也可以腾空飞翔，因为他们有爱情的力量。而对于自幼体弱多病、却对一切感到惊奇的雅克-昂利·拉蒂克而言，“悬浮”——即便只是摄影机才能捕捉的瞬间悬浮，也能成为实现内心的超越和幻想的寄托所在。同样，受达利“原子时期”绘画的启发，

左：中国古代的“飞天”　　中：贝拉·夏加尔《恋人》

1 龚云表：《羽人》，载《飞行的诗意》（徐累主编），北京：中国人民大学出版社，2009年版，第34页。
2 丹尼尔·马什舍：《夏加尔——醉心梦幻意象的画家》（周梦罴译），上海世纪出版集团译文出版社，2003年版，第47页。

上：波提切利《维纳斯的诞生》局部
下左、下中：雅克 - 昂利 · 拉蒂克摄影中的身体“悬浮”
下右：菲利普 · 豪斯曼《原子达利》

摄影家菲利普 · 豪斯曼（Philippe Halsman）拍摄的《原子达利》（*Dali Atomicus*）以及此后的跳跃学（Jumpology）摄影作品也将瞬间“悬浮”的诗意表现得淋漓尽致。

一定意义上，拉蒂克和豪斯曼摄影作品中的瞬间“悬浮”揭示的正是重与轻、挣脱与吸引、自由与束缚的悖论。相比之下，迪勒和斯科菲迪奥(Elizabeth Diller and Ricardo Scofidio)为2002年瑞士博览会设计的《模糊建筑》(The Blur Building)则试图通过喷雾装置将建筑转化为一朵“悬浮”在水面之上的“巨云”，从而在无需克服重力的情况下“巧妙”地为“悬浮”定格。但是很显然，要想表现“悬浮“的诗意，建筑学既不能像拉蒂克和豪斯曼那样通过拍摄将瞬间定格，也不能像迪勒和斯科菲迪奥依靠喷雾装置制造“悬浮”的幻象。真正的“建筑悬浮”需要在建筑结构的层面解决更为复杂的问题。

这是霍尔的万科中心需要面临和解答的问题，也曾是勒 · 柯布西耶的萨伏伊别墅和库哈斯的波尔多住宅面临和致力于解答的问题。关于萨伏伊别墅，柯布自己这样写道：“基地：一片大草坪，中间微微隆起。……整个房子就是一个悬浮在空中的盒子，…… 它的主人们，出于对田园风光和美丽的乡间生活的向往，来此居住，站在空中花园，透过四周的条形窗，面对眼前这片未受侵扰的土地浮想联翩。他们的居家生活宛如沉浸在一个维吉尔

迪勒和斯科菲迪奥　喷雾装置与“模糊建筑”

勒·柯布西耶草图中的萨伏伊别墅

般的梦境之中。”[1] 这就是人们习惯将萨伏伊别墅解释为“一个居住的机器、一种体量与空间穿插的组合、一种时空流溢”之时，柯林·罗（Colin Rowe）在《理想别墅的数学》中试图揭示的蕴含在该建筑中的另一番诗情画意。只是，柯林·罗的关注点似乎仅在于萨伏伊别墅置身其中的田园风光，而未曾意识到“悬浮”在这一“维吉尔田园之梦”中的意义。在这里，勒·柯布西耶实现“悬浮”的手段是“新建筑五点”中使底层架空的独立支柱。更有意思的是柯布的草图，不仅地面被刻意弯曲一样，而且建筑的独立支柱也被刻意细化，以突显上面那个方盒子的“悬浮”之感。

萨伏伊别墅是一个现代建筑的“悬浮之梦”。它的结构体系是稳定的，恰如柯布的建筑体系是稳定的一样。但是，七十多年后，在库哈斯的波尔多住宅（Maison à Bordeaux）中，“悬浮”这个对于房屋的主人——一位由于车祸而身陷轮椅的作家——来说或许不无拉蒂克式的超越和浪漫[2]的建筑概念，经过结构工程师塞西尔·巴尔蒙德（Cecil Balmond）的努力，俨然成为探索建筑学“异规”（informal）——其中无疑也包括结构系统的“异规”——的一次创造性的尝试。[3]

“漂浮的地平线——一个位于最大化景观园林之上的水平向超高层建筑”是霍尔在万科中心设计伊始就提出的概念。一定意义上，正是这个概念将“水平线性”“生态”和“悬浮”等过去几年霍尔事务所中国项目的设计理念集于一身。一方面，如同包括南宁柳沙半岛方案在内的其他项目，这个“漂浮的地平线”以不规则水平线性的方式延展，不同

1 转引自 Colin Rowe, “The Mathematics of the Ideal Villa,” *The Mathematics of the Ideal Villa and Other Essays* (Cambridge, Massachusetts and London: The MIT Press, 1973), p.2.

2 关于波尔多住宅，这位身陷轮椅的作家这样说道：“与人们希望的正相反，我并不想要一个简单的住宅。我要一个复杂的住宅，因为这个住宅将定义我的世界。”见 Terence Riley, The Un-Private House (New York: The Museum of Modern Art, 1999), p.92.

3 见塞西尔·巴尔蒙德：《异规》（李寒松译），北京：中国建筑工业出版社，2008 年版，第 21-44 页。

塞西尔·巴蒙《异规》英文版

之处在于现在它不仅是一个“地平线”，而且也是一个水平的摩天楼，一个平放的帝国大厦。据霍尔自己说，之所以与帝国大厦相提并论，是因为这是一个他在自己的纽约事务所每天都能看到的建筑。但是，面对摩天楼俨然已经成为一种“常规”的当今中国建筑现状，人们也许不难辨别霍尔通过帝国大厦的比拟试图提出一种摩天楼“新范式”的野心。另一方面，与现代 MOMA 和佛手湖建筑艺术博物馆的局部“悬浮”不同，万科中心的“悬浮”主题在概念上更为彻底，它是整个建筑的“悬浮”。而且，帝国大厦的比拟足以说明，这个“漂浮的地平线”的巨大尺度已经远非萨伏伊别墅或波尔多住宅所能比拟。

通常，巨大的水平线性布局势必会增加横向交通的压力。好在万科中心是由万科总部、会展中心、酒店、SOHO 公寓等相对独立的功能组合而成的，彼此之间没有频繁进行水平联系的要求。与此同时，尺度巨大的“悬浮”无疑对结构设计提出了强有力的挑战。在这方面，万科中心的“悬浮”是通过一个集大跨、钢结构、悬拉索和预应力为一体的混合结构予以实现的。根据建筑师概念方案的结构设想，结构工程师提出了“斜拉桥上盖房子”的结构理念，它在钢筋混凝土核心筒、实腹墙（包括包含在局部地面建筑形成的“土丘”之中的实腹墙）和立柱的配合下，

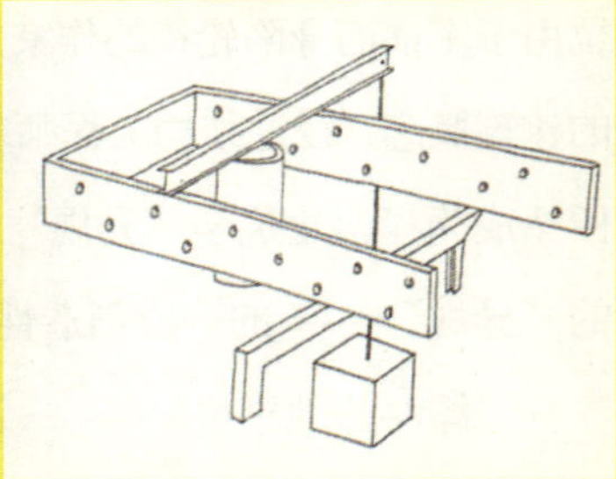

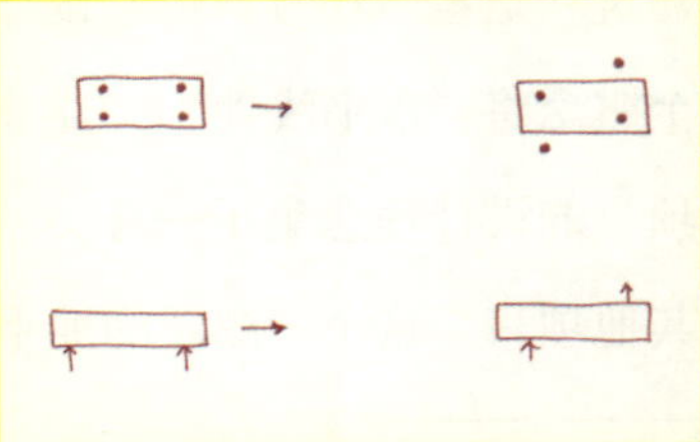

左上：波尔多住宅　　右上：波尔多住宅结构图解
右下：从稳定结构系统到动态结构系统

通过高强钢丝斜拉索将结构荷载传到基础。与此同时，为减少造价，设计者只在建筑主体的底部采用钢结构楼盖，而在其余楼层采用混凝土结构。这要求设计者在结构计算和施工方案中力求充分发挥首层钢楼盖及顶层混凝土楼盖的轴向刚度和承载力，寻求预应力值的最优化。另外，人们也不难注意到，为打破垂直柱的单调感，在“悬浮”中起支撑作用的群柱之中有几根被特别做了倾斜化处理。

因此，在突破多米诺住宅（Maison Dom-ino）所代表的框架体系的同时，万科中心的结构并没有像波尔多住宅那样寻求一个刻意的不稳定系统。而且，与波尔多住宅相比，万科中心“斜拉桥上盖房子”的结构理念也没有得到直接和强烈的外部表现，而是隐约呈现在玻璃幕墙和金属遮阳板的背后，只有在夜晚室内灯光的衬托下才变得比较明显。当然，这种比较有些像柯林 · 罗在比较勒 · 柯布西耶的加歇别墅和格罗皮乌斯的包豪斯校舍时所说的，是在两个大小相差悬殊的建筑之间进行比较，因而并不公允，尤其是对尺度较大的建筑而言。[1] 但是，如果将万科中心与同样由库哈斯和巴尔蒙德联袂设计、也同样旨在创立摩天楼新范式的 CCTV 进行比较的话，我们却不难看出，万科中心的结构体系基本还是实用性和表现性兼而有之的，而 CCTV 则似乎由于造型概念的先天不足而陷入了一种不仅使结构造价变得十分昂贵，而且也使实现这一造型的结构手段变得没有多少表现性可言（玻璃幕墙上呈不均匀变化的斜交网格应该除外）的尴尬境地。

不过，由于尺度巨大，万科中心“建筑悬浮”的整体感其实很有限。

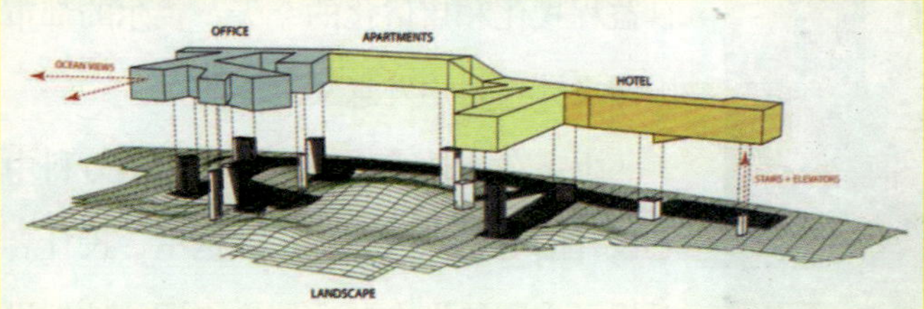

左上：南京佛手湖美术馆
右上：纽约帝国大厦高度与深圳万科中心长度的比较
下：万科中心的“悬浮”

1 Colin Rowe and Robert Slutzky: “Transparency: Literai and Phenomenal”, *The Mathematics of the Ideal Villa and Other Essays*, p.171.

这个“水平向超高层建筑”更像一个底部架空而非“悬浮”的建筑。在这里，感受“悬浮”的最好位置都是局部性的，比如站在建筑主体的下方，让这部分建筑主体变成观者头顶的华盖，再看到中景和远景中的建筑局部，特别是那些巨大的端部悬挑。这时，如果能够注意到一两处吊挂式的逃生楼梯和以吊挂的玻璃盒子面貌出现的“深圳之窗”，“建筑悬浮”之感还是不容否认。相对而言，夜晚的“悬浮”感更加美轮美奂，因为在泛光灯的局部照射之下，白天相对沉重的建筑主体得到弱化，建筑底部的彩色吊顶更加突显出来，而核心筒磨砂玻璃后面的灯光则使这个支撑元素更加非物质化了。

一定意义上，“悬浮”是“锚固”的对立面。但是，正如霍尔早在《锚固》中已经表述的：“每一个建筑都有一个场地。人们由此得到建筑的意图。自有建筑开始，建筑与场地就相互依存。过去，通过使用地方材料和工艺以及景观与历史、神话事件的关联，这一联系被无意识地呈现。今天，应该用新的方法寻找场地与建筑的联系，它是现代生活建设性转变的组成部分。”[1]

确实，除了表现“悬浮”的诗意，万科中心底层架空的最大好处恰恰在于对整个场地的释放。一方面，它力图使场地达到最大的公共化，进而将整个场地转化为一个对市民开放的城市公共空间。在当代中国城市中各自为政、他人不得入内的“单位”格局中，这样的努力不仅使万科中心获得全然不同的城市意义，而且毫无疑问，需要甲方的胆识、勇气和认同。另一方面，场地的释放也为一体化的景观设计提供了条件。值得一提的是，这是一个由霍尔事务所自己承担的景观设计。受 20 世纪优秀的巴西景观设计师、生态学家和自然主义者罗贝托 · 布尔勒 · 马克斯（Roberto Burle Marx）的影响，霍尔万科中心的景观设计强调原生态和田园式的质朴，避免过多色彩的搭配（这一点也可解释为彩色烤漆板吊顶带来的必然要求），最大限度地挖掘少量植物品种的景观潜力。应该说，在中国的景观设计日益程式化和庸俗化的今天，霍尔的都市田园设计不无启发意义。

CCTV

生态节能技术的开发和运用是万科中心设计的重要内容之一。在这里，设计者采用了太阳能发电、冰蓄冷空调、地板送风、新风热回收、智能照明控制、智能调

1 Steven Holl, *Anchoring* (New York: Princeton Architectural Press, 1989), p.9.

斯蒂芬·霍尔 李虎 深圳万科中心

节遮阳板、雨水收集和水循环处理系统等诸多主动和被动的生态节能措施。必须承认，其中有些措施如太阳能发电如果就其使用规模而言即使在国内也不算领先，但是如此综合地考虑节能措施应该说还是开创性的。当然，如同一切新尝试一样，这些措施是否真能达到设计者所设想的节能效果还有待观察，其经验也有待总结。

但是，最终赋予万科中心某种都会田园般表现性的显然还是霍尔的景观设计。面对这样一个集建筑、景观和生态技术为一体的作品，人们有理由期待，在不久的将来（因为现在还不是这样），当人们来到这片都市田园，行走在这个巨大的蜿蜒曲折的“建筑悬浮”之下，沐浴徐徐吹来的海风，眼看色彩缤纷的铝板吊顶和核心筒磨砂玻璃的光影反射以及智能调节产生的金属遮阳板的微妙变化，耳听草丛中的蛙声，鼻嗅植物的气味，一种建筑的“现象学”全体验将在不知不觉中悄然而至。

如果说柯布的萨伏伊别墅是现代建筑“维吉尔田园之梦”历史意识的一次呼唤，那么霍尔的万科中心则是当代建筑“基于爱的建造”的一首都会田园之歌。更为重要的是，正是后者使我们看到，即便在我们这个玩世不恭的时代，我们仍然应该期待这样一种建筑的出现：它渴望建筑的诗性表现、又不回避建筑的伦理责任，尽管两者之间从来不存在必然的对等关系；它致力于形式创造、又不放弃建筑改善社会的可能和努力，尽管对于每一个单体项目而言，这种可能常常是有限的；它在希望建筑好玩有趣的同时，又要谨防使建筑学仅仅成为自娱自乐的游戏，尽管历史地看，自娱自乐也是学科发展的必要前提之一。

结语

“建筑现象学”不仅是材质、光、色彩、触感、知觉的建筑学，而且也是“慢”的建筑学。它给我们一片宁静的空间，让我们能够在纷乱和匆忙之中停下来思考问题，回溯本源，追寻建筑的真谛。刚出道的霍尔曾经这样做过，没有多少设计任务、只能做小型项目或者研究性项目的霍尔也曾这样做过，但是一个已经大红大紫、有太多项目的霍尔还能够继续这样做吗？还有，随着霍尔效应的扩大，反过来又形成市场对霍尔建筑花样翻新的期待和要求，“知觉的问题”会失去其最初的“现象学”意义，转而成为“奇观建筑学”的帮凶吗？对此，我们现在还不得而知。

就万科中心而言，我们应该说万科是幸运的，因为它在数轮不尽如人意的方案设计之后，最终选择了霍尔和他的方案，其结果我们已经看到了 —— 一个能够驾驭大型项目的霍尔所能完成的好的作品（当然不容否认的是，在这样的过程中，霍尔的中方合伙人李虎的作用也至关重要）；但是，在这段缘分之中，更为幸运的也许是霍尔，因为至少在中国，他邂逅了万科这样一个甲方，一个致力于建设现代企业文化、勇于探索建筑的公共意识和可持续发展之路、同时又是一个具有丰富开发建设经验和建筑实施能力的甲方，其结果我们也看到了 —— 一个不仅在完成度上远远好于霍尔自己在中国的其他项目，而且从设计理念到施工质量上来说或许也是近年来国际知名建筑师在中国建成的完成度最好的建筑。“晚期资本主义‘景象社会’的实质已经充分表明，开明的业主是取得一切具有文化意义的建筑成就的前提”，肯尼斯·弗兰姆普敦（Kenneth Frampton）在《建构文化研究》（*Studies in Tectonic Culture*）中如是说。[1] 霍尔的万科中心似乎再一次验证了弗氏的这一论断，尽管是在看似不尽相同的社会和文化语境之中。

1 肯尼斯·弗兰姆普敦：《建构文化研究——论 19 世纪和 20 世纪建筑中的建造诗学》（王骏阳译），北京：中国建筑工业出版社，2007 年版，第 392 页。

7

意义的探究

——克里斯蒂安·诺伯格-舒尔茨建筑理论评述[1]

1 最初发表于《世界建筑》1997年第4期（总94期），收入本文集时有修改。

说到北欧现代建筑，人们似乎就不能不提及挪威当代建筑历史和理论学家克里斯蒂安·诺伯格-舒尔茨（Christian Norberg-Schulz）和他的建筑理论。事实上，由丹麦学者尼尔斯-乌勒·隆德（Nils-Ole Lund）撰写、现已成为有关北欧现代建筑方面经典著作的《北欧建筑》（*Nordisk arkitektur*）一书就为论述诺伯格-舒尔茨的建筑理论专辟一章。隆德这样做不是没有道理的。自从 1949 年在瑞士苏黎世高工（ETH）修完建筑学科之后，诺伯格-舒尔茨就以各种形式活跃在挪威、北欧以及世界建筑界。他先是从事建筑实践，并组建和积极参与世界建筑师协会 CIAM 挪威分会 PAGON 的活动。其后于 1963 年出任挪威建筑杂志《建筑艺术》（*Byggekunst*）主编 15 年，是该杂志任期最长的一位主编。这期间，他为该杂志撰写了大量专题文章，对挪威建筑界影响甚大。此外，从 20 世纪 50 年代初开始，诺伯格-舒尔茨就担任部分建筑教学工作。70 年代中旬被任命为奥斯陆建筑学院教授之后，更以著书育人为己任，锲而不舍。当然，在诺伯格-舒尔茨多样化的建筑生涯之中，最富成果并给他带来广泛国际声誉的还是他大量的历史和理论著作。在这一点上，诺伯格-舒尔茨最像他在苏黎世高工的老师吉迪恩（Sigfried Giedion）；对巴洛克的热爱和对 20 世纪的关注兼而有之；历史研究和理论著说齐头并进，相辅相成。

不过，被普遍视为诺伯格-舒尔茨理论生涯成名之作的《建筑的意向》（*Intentions in Architecture*，以下简称《意向》）一书却不是一部吉迪恩式的著作。因为如果说吉迪恩的著作主要是以传统艺术史的观点和方法写成的话，那么《意向》一书要做的正是摆脱传统艺术史的范畴和概念而另辟蹊径。早在 20 世纪 50 年代初期获美国福布莱特（Fulbright）基金会奖学金赴哈佛大学进修期间，诺伯格-舒尔茨就深受同宿舍的挪威老乡、心理学家乌拉夫·斯戈道尔（Olav Skådal）的影响而对格式塔心理学产生了浓厚的兴趣。同时他广泛接触社会学、符号学等学科的内容。此后

左：尼尔斯-乌勒·隆德《北欧建筑》丹麦文版　中：尼尔斯-乌勒·隆德《北欧建筑》英文版
右：克里斯蒂安·诺伯格-舒尔茨《建筑的意向》英文版

正是对这些学科的应用构成了《意向》一书的理论框架。这是一本雄心勃勃的著作。它试图从整体上把握建筑的多重维度。它不仅涉及了建筑历史、美学、技术、教育等传统问题，更涉及了与社会学、语言学、感知心理学等学科相关的新问题。它试图建立一套科学、客观的语言来描述和评价与既定建筑任务相应的形式。

《意向》一书于1963年以英文出版后，以其综合严谨的特点给当时正处于无所适从的国际建筑界带来了一阵清风，并很快被译成多种文字。然而，就是这样一部雄心勃勃的著作，并没有能够如其作者的初衷真正排除主观性而使建筑成为“客观”和“科学”的。相反，它从一开始就以其书题道出了一个充满主观性的概念：意向。许多年以后，诺伯格-舒尔茨自己这样写道：“在《建筑的意向》一书中，建筑和艺术是以从自然科学中借鉴的方法来‘科学’地分析的。我现在并不认为这样的做法完全有误，但我觉得其他的方法更富有启发性。”[1] 诺伯格-舒尔茨这里说的其他方法就是他后期著作中所谓的“现象学”方法。

20世纪的现象学多是与胡塞尔（Edmund Husser）、海德格尔（Martin Heidegger）、梅洛-庞蒂（Maurice Merleau-Ponty）等大师的名字以及他们的学说连在一起的。事实上，诺伯格-舒尔茨自己就在其著作中频繁提及胡塞尔“面向事物本身”（Zu den Sachen selbst）的口号和“生活世界”（Lebenswelt）的概念，也不止一次地引用梅洛-庞蒂感知现象学的观点，更把海德格尔的学说视为自己建筑理论的基准。但是正如有些学者指出的，诺伯格-舒尔茨的建筑现象学实际上是远离了从胡塞尔到梅洛-庞蒂意义上的现象学的。充其量它只是一种以视觉形式主义为基准对应于建筑和自然景观及其相互关系的“现象学”。[2]

左：胡塞尔《欧洲科学危机与超验现象学》中文版　左中：胡塞尔《生活世界现象学》中文版
右中：赫伯特·斯皮伯格《现象学运动》中文版　右：安德烈·罗宾耐《模糊暧昧的哲学——梅洛-庞蒂传》中文版

1 Christian Norberg-Schulz, *Genius Loci - Towards a Phenomenology of Architecture* (New York: Rizzoli, 1984), p.5.

2 见 Roderick J. Lawrence, “Architecture of the City Reinterpreted: a Critical View,” *Design Studies*, Vol.6 No.3, July 1985.

我个人认为这一批评是有一定中肯性的。诚然，诺伯格 - 舒尔茨谈的是建筑的现象学，他不可能也没必要像胡塞尔那样为澄清世界和意识的本质结构而进行“本质还原”和“先验还原”；或者像梅洛 - 庞蒂那样为把“肉体”从被近代形而上学遗忘的角落中拯救出来，并使其重获全新的、不同于“思我”(Cogito) 的主体性意义而在客观和主观之间走钢丝。[1] 但是，从一开始现象学与其说是一种方法毋宁说是一种精神。它力排不加反省偶然拾来的概念框架，这些框架原为探索事物本质而设，后来却导致了浮夸的结构和空洞的话语而阻碍人们探入事物本质的眼光。它唾弃体系哲学，以便自下而上，而不是自上而下地进行“生活世界”之反思和解释，无论这一“生活世界”在不同学科中的具体内容多么不一样。[2] 诺伯格 - 舒尔茨以“现象学”为基础的建筑理论是否具有这样的精神，这不是没有疑问的。

有一点却是肯定的：诺伯格 - 舒尔茨通过他的“现象学”转向至少从字面上认同了西方 20 世纪以来在“理性的批判”向“文化的批判”的过渡中超越“逻各斯中心主义”而意欲深究“逻辑背后”事物真正本原的思想。[3] 要证明这一点，人们只需看一看诺伯格 - 舒尔茨放在他建筑现象学著作最首要部分的卡夫卡（Franz Kafka）的这段话：“逻辑无疑是不可动摇的，但它却无法阻挡人对生活的向往。”[4]

其实，还是在《意向》之后不久发表的一篇题为“建筑的意义”一文中，诺伯格 - 舒尔茨就一再强调建筑应属于“生活的世界”而非“逻辑的世界”。他因此批评西方文化自文艺复兴以后对现实理解日趋数量化和抽象化的倾向。这种倾向也浸漫到建筑中来，其结果据诺伯格 - 舒尔茨看就是 20 世纪的功能主义建筑。他指出：一方面，功能主义将建筑的功能数量化、物质化；另一方面，技术理性的进步虽然使自由流通空间成为可能，但并没有重温巴洛克空间的“超自然性”。相反，它使建筑本身沦为统一性和逻辑性的牺牲品，而无法成为“富有意义的形式”。[5]

诺伯格 - 舒尔茨的以上说法在他后来的历史著作《西方建筑的意义》(*Meaning in Western Architecture*) 中得到了部分修正。因为在这部意欲重写西方建筑史的著作中，西方历史上不同文化时期的建筑，无论是古希腊的还是中世纪的或文艺复兴的，均被视为那个特定时期宗教和哲学主导思想的物质表现，并因此成为表达和承载人对世界和自己的存在意义

---

1 关于胡塞尔和梅洛 - 庞蒂对现象学的贡献，参见 Herbert Spielberg, *The Phenomenological Movement*（The Hague, Boston, London: Martinus Nijhoff Publishers, 1982）。

2 倪梁康：《现象学精神：为何？何为？》，载《读书》1995 年第 10 期。

3 甘阳：《从“理性的批判”到“文化的批判”》，载《读书》1987 年第 7 期。

4 Christian Norberg-Schulz, *Genius Loci*, p.5.

5 Christian Noberg-Schulz, “Meaning in Architecture” in *Meaning in Architecture*, ed. George Baird and Charles Jencks (London: The Cresset Press 1969), pp.214-229.

左：诺伯格-舒尔茨《西方建筑的意义》英文版
右：诺伯格-舒尔茨《西方建筑的意义》中文版

之理解的象征形式。就这一点来说，功能主义也不例外。诺伯格-舒尔茨强调指出："如果以为功能主义仅仅对效率感兴趣的话，那就错了。与历史上任何一个伟大运动一样，功能主义关注的首先是意义，即为人提供存在立足点的问题。"[1]

从这较后的观点来看，诺伯格-舒尔茨对功能主义的不满与其说是指向功能主义建筑优秀作品中理性世界观的偏差，毋宁说是针对20世纪，尤其是二战之后以功能主义为名而粗制滥造，缺少个性的大量性建筑的问题。一向主张整体环境（total environment）的诺伯格-舒尔茨自然是不能仅仅满足于少数优秀作品的。也正是在整体环境的意义上诺伯格-舒尔茨提出了他后期著作中一再使用的"环境危机"（environmental crisis）的说法。所谓"环境危机"首先是他早在《意向》一书中提出的现代环境视觉混乱（visual chaos）的问题。在诺氏看来，意义就是关系，它只能在和谐的视觉关系中存在。视觉混乱中不具意义。继《意向》之后，诺伯格-舒尔茨又提出诸如传统围合空间的失落、文化个性的丧失，单调雷同的城市环境等问题。不消说，诺伯格-舒尔茨并不是注意到这些问题的第一人，也不是针对这些问题而提出"场所"（place）概念的唯一学者。[2] 但是他由此建立的建筑理论，亦称"场所精神"（genius loci），却可以说是与众不同的。

也许，对于素有"风水"概念的中国人来说，诺伯格-舒尔茨的"场所精神"说并不难理解。与"风水"理论相似，"场所精神"也是一个古老的概念。据诺伯格-舒尔茨说，它源自古罗马人对场所守护神的信仰。按照这一信仰，场所不分大小，均有自己的守护神。是守护神赋予场所

1 Christian Norberg-Schulz, *Meaning in Western Architecture* (London: Studio Vista, 1980), p.202.

2 关于这个问题的讨论，参见 Edward Relph, *Place and Placenessless* (London: Pion Limited, 1976).

诺伯格 - 舒尔茨
《场所精神——走向建筑现象学》英文版

以及生活在该场所的人们生命和活力，伴随他们从生到死并决定他们的特征和本质。[1] 但是，如果说“风水”理论最终沦为一种以诸如去邪辟祸、招财进宝之目的的功利性手段的话，诺伯格 - 舒尔茨对“场所精神”概念的重新挖掘则更注重西方自古以来人对自然理解的神化或者说隐喻思维。此外，如果说中国文化中带有公共性质的“场所”更多是一个世俗的集市世界（夫子庙、灯会）的话，那么诺伯格 - 舒尔茨的“场所”理论则似乎反倒让我们通常认为过分“注重物质”的西方文化的精神维度充分展现出来。他因此追溯诸如山峰作为天地联姻表现的说法，重温基督教将光视为知识和爱之象征的概念，以及文艺复兴、浪漫主义和象征主义绘画所表达的自然主题。诺伯格 - 舒尔茨强调，意义就是关系，人造场所的存在意义最终就得在与自然场所的关系之中寻找。

因此，《场所精神——走向建筑现象学》（*Genius Loci: Towards a Phenomenology of Architecture*）一书不仅试图用光、质理、围合的程度、地面起伏的状态等具体的现象特征范畴来分析和描述自然和人造场所，更致力于二者的关系。首先，诺伯格 - 舒尔茨将西方南至地中海地区、北到斯堪的纳维亚半岛的自然环境分为宇穹性的（cosmic）古典性的（classic），和浪漫性的（romantic）等原型。此外，鉴于原型和原型之间不能截然分开而必有过渡，他也提出复杂性的（complex）原型说。诺

太保相宅图

1 Christian Norberg-Schulz, *Genius Loci*, p.18.

诺伯格 - 舒尔茨
左上："宇穹型"建筑
左下："古典型"建筑
右上："浪漫型"建筑
右中和右下："复合型"建筑

伯格 - 舒尔茨坚持认为在西方建筑史漫长的演变发展中产生的丰富多彩的形式和风格在很大程度上与既定的自然场所之精神有关。于是，他根据以上自然环境的原型，也归纳了相应的几种建筑原型，称阿拉伯地区的建筑为"宇穹型"的，古希腊建筑为"古典型"的，斯堪的纳维亚地区最初的木质建筑为"浪漫型"的，而诸如哥特教堂和巴洛克花园宫廷建筑则为"复杂型"的。当然，诺伯格 - 舒尔茨并没有止步于这种原型性的理论。为了进一步说明自己的观点，他还重点考察了布拉格、喀土穆（Khartoum，苏丹首都）和罗马这三个城市及其与所在自然环境的关系。诺伯格 - 舒尔茨力图说明的是，即使城市这样更为高级和复杂的人造场所，其组织结构和建筑总体也表现了"人对自然以及自己总的生存状态的理解"，因而是富有存在意义的。[1]

如此简要地概括"场所精神"说，诺伯格 - 舒尔茨一定会不以为然。确实，《场所精神》一书图文并茂，无论是诺氏对远古神思的追溯，还是他对特定环境的分析描述，其丰富性和生动性均不是几个条条杠杠所能替代的。更具诱惑力的是诺伯格 - 舒尔茨对海德格尔哲学，尤其是晚期哲学的解释和应用。一般接触过海德格尔哲学的人恐怕很少有不为其晦涩朦胧而望而却步的。那些生僻的字眼，拗口的句子，闪闪烁烁时隐时现的思路，常常叫人大有不堪卒读之感。然而，经过诺伯格 - 舒尔茨栩栩如生、化繁为简的注说，似即柳暗花明，另有一番天地展现。这包括海氏早期关于此在（Dasein）在世界中的存在结构及其空间性的理论，晚期对物之物性，艺术品的起源和本质，语言作为存在之家园，思、居、

1 Christian Norberg-Schulz, *Genius Loci*, p.50.

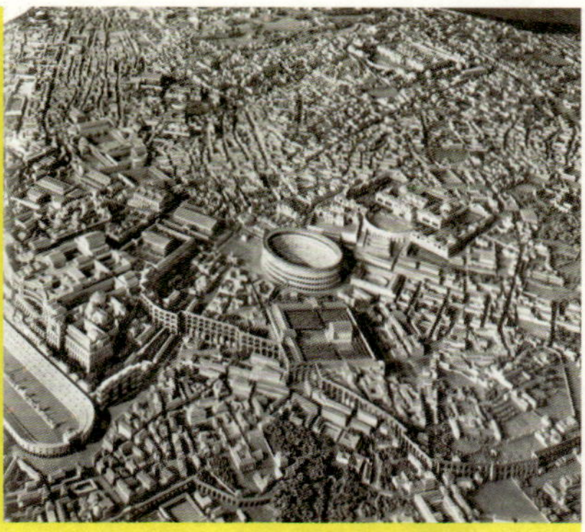

左上：布拉格　　右上：罗马　　下：喀土穆

筑三者的内在关系，以及人的诗性栖居等问题的思考。据此，诺伯格-舒尔茨不仅为他所关切的建筑的“存在意义”找到了哲学基础，也使人们看到他似乎更多的是在海德格尔现象学的意义上来展开他的“建筑现象学”的。须知，海德格尔与胡塞尔在现象学观念上的分歧之深以致他后来为避免同一名称引起的混乱而放弃了“现象学”这个术语。[1]

但是，诺伯格-舒尔茨这样做不是没有问题的。一方面，建筑的“通俗”注说对于一种哲学理论来说常常只见大树不见森林而导致误解，这一点在诺伯格-舒尔茨那里似未能免。例如，挪威学者安芬 · 波-芮格（Arnfinn Bø-Rygg）曾经指出诺伯格-舒尔茨将海德格尔早期对存在结构的分析与其晚期对语言的思考混为一谈，而海德格尔自己则在他所谓的“转向”（die Kehre）之后曾表示这样的结构是传统行而上学的问题而应予以克服（其巨著《存在与时间》的第二卷因此从未完成？）[2]。诸如此类的问题也许过于哲学性而非本文篇幅所能讨论。笔者以为，倒是一些诺伯格-舒尔茨建筑理论中与海德格尔思想较为一致的地方更有必要多谈几句。

一般认为，晚期的海德格尔不再试图勾画此在的整体结果，而将守护存在的状态视

海德格尔《存在与时间》中文版

1 参见 Spielberg, *Phenemenological Movement* 有关海德格尔的章节 . 也见陈嘉映，《海德格尔哲学概论》，北京：生活 · 读书 · 新知三联书店，1995 年版，第 49-55 页（第二章第四节）。

2 Arnfinn Bø-Rygg, “Arkitektur og filosofi. Om Christian Norberg-Schulz' bruk av Heideggers senfilosofi”, *Norsk filosofisk tidsskrift*, 1986-1/2, pp.37-53(40).

左：海德格尔《诗，语言，思想》英文版　中：海德格尔《人诗意地安居》中文版
右：海德格尔《荷尔德林诗的阐释》中文版

为“思”的任务。为此，他的方法依然是现象学的（当然是他所理解的现象学），即通过寻找存在显现的住所来将其去敝。在海德格尔看来，艺术作品，语言都是这样的住所。[1] 其中尤其重要的是语言，而语言又和诗紧密相连。这不仅因为诗的活动领域是语言，更因为语言作为存在的显现首先表现为诗。“诗是给存在的第一次命名，是给一切存在物的第一次命名。”这与其说是诗把语言当作手边的原始材料来运用，不如说是诗使语言成为可能。“命名不是贴标签，不是使用语词，而是唤入言词。命名呼唤。”呼唤需要敞开和接纳，而诗作为敞开者就“意味着某些无阻碍的东西。它之所以无阻碍，是因为它不设置界限，它之所以不设置界限，是因为它本身无任何界限，它是一切无界限的东西的整体。”这一点使诗在贫瘠时代的重要性更为突出。“语言是存在的家”，语言的贫瘠就是人的存在的贫瘠。克服贫瘠的时代还需诗的呼唤。当然，海德格尔这里所谓的诗并不完全指文字性的。“一切艺术本质上都是诗”。又曰：“人诗意地安居。”[2]

黑森林小屋

这样看来，海德格尔的晚期哲学不仅呼吁诗化，而且也颂扬开敞的艺术了。值得注意的是，无论说诗、谈艺术还是讨论物性或栖居，海氏的观点都是有一定

1　陈嘉映，《海德格尔哲学概论》，第 55 页。

2　Martin Heidegger, “What Are Poets For?”, “Language”, “The Origin of the Work of Art”, 及 “... Poetically Man Dwells ...”, *Poetry, Language, Thought* (New York: Harper Perrenial, 1971).

内涵的。首先，受诗人荷尔德林（Friedrich Hölderlin）对神话的领会的启发，“四大”，亦即天、地、人、神，是海氏晚期反复使用的字眼。物之物性就在于聚集“四大”，而诗人则在诸神遁走的贫瘠时代截取神的声音并揭示存在。其次，虽然“四大”合为世界，海德格尔却赋予土地特别的地位。在一定程度上，土地概念可以说是他早期著作《存在与时间》中自然概念的具体化。不过，海德格尔对土地的厚爱另有一番曾被部分学者称为“农耕保守主义”（conservativisme agraire）的原因。[1] 出于这一原因海氏不仅把荷尔德林《还乡》一诗解释为家园与天命二者的联系及其根本性，更把海贝尔（Johann Peter Hebel）视为家乡诗人的典范。至于海德格尔之所以用梵高画的一双鞋来说明艺术作品的起源，倒是因为那是一双农妇的鞋。从农妇到土地，从土地到“四大”，似乎画一双别的什么鞋就无法成为真正的艺术作品。正如波 - 芮格指出的那样，海氏的“农耕保守主义”不仅使他将 20 世纪的艺术（尤其是抽象艺术）多视为堕落的甚至具有破坏性的，也使他的许多思想失去与我们时代的文化景观，尤其是城市景观的相关性。在海德格尔的眼里，现代城市正是人无根基存在的典型。[2]“哲学是本真的思乡病”[3]，遵循海氏的思想，人的本真存在唯有通过返归类似他的“黑森林”小屋这样的家园才能思考和获得。

梵高：《鞋》（油画）

诺伯格 - 舒尔茨倒没有走这么远。他不鄙视现代艺术。相反，他用现代抽象艺术作品来阐述海氏的“物”之概念。[4] 此外，如我们前面已经看到的，他也不拒斥城市。相反，他把布拉格、罗马和喀土穆这样的城市视为“人诗意地栖居”的优秀范例。然而，且不说这一观点中人工场所与自然场所一对一关系的可疑性，即便诺伯格 - 舒尔茨的分析本身完全正确，现代城市的发展也离这些范例相去甚远。纽约、东京、上海，我们如何再能看到这些城市的结构和形象具有与它们所在自然场所相一

1 Arnfinn Bø-Rygg, “Arkitektur og filosofi. Om Christian Norberg-Schulz' bruk av Heideggers senfilosofi”, Bø-Rygg, footnote 16.

2 Ibid, p.48.

3 Mark Wigley, “Heidegger's House: the Violence of the Domestic” in *Columbia Documents*, 1992-1.

4 Christian Norberg-Schulz, “The New Tradition Today” in *International Laboratory of Architecture and Urban Design*, Year Book, 1982, 52-57, 以及 *The Concept of Dwelling*, (New York: Electa/Rizzoli, 1985).

当代城市

致的整体意义和精神？大量性建造的住宅区、机场、工业区、高速公路，我们如何再能以“物”的概念去聚集“四大”？不奇怪的是，诺伯格-舒尔茨对现代环境多有批评。批评之一就是我们前面已经提过的“环境危机”说。对于这一学说，如果是从现代生态学的角度来看，我倒能同意诺伯格-舒尔茨的观点。需要指出的是，他理论的重点其实在于建筑、城市的空间和形式以及它们的“存在意义”。

换言之，诺伯格-舒尔茨建筑理论的问题不在于它提出了建筑作为表达“场所精神”的可能性甚至必要性——在这一点上，笔者完全同意诺伯格-舒尔茨对建筑事务所 MLTW 设计的加州海边住宅的评价和赞誉，而在于它将城市视为表达他所谓“存在意义”的整体艺术。在后一个方面，面对 20 世纪以来的城市现实，诺伯尔-舒尔茨理论的有效性几乎为零。

另一方面，诺伯格-舒尔茨的理论又把物质意义上的问题等同于现代社会人之“异化”而持一种类似物质形态决定论的观点。诺伯格-舒尔茨曾断言：“我们有理由相信人类的异化今天如此普遍，在很大程度上取决于现代环境提供了太少定向和认同的可能性。”[1] 但是，如果事情真如诺伯格-舒尔茨所说，那么如何解释正是在布拉格这座“为人的存在提供立足点”的典范性城市里，卡夫卡写下了大量表达他异化感情的文学著作这一事实呢？反之，在所谓“视觉混乱”或被诺伯格-舒尔茨等学者称为“非场所”（placeless 或 non-place）的现代环境里，认同感和根基感是否

1 Christian Norberg-Schulz, *Genius Loci*, p.180.

MLTW　加州海边住宅

便不可能？欧洲城市的郊区多有移民和低收入者的聚居区。这些居住区多是二次大战后大量性住房建造的产物。建筑形象单调，缺少个性，也没有传统城市的格局，是典型的"非场所"。常有新闻媒介对此作报道，把它们作为"问题成堆地区"来讨论。然而，正如瑞典学者本格松（Jan Bengtsson）指出的，每有这样的报道，就每有当地居民倍感对自己的侮辱。原因是，这些居民已经认同于自己的"生活世界"，以致对后者的攻击便被视为对前者的攻击。[1]不是说这些地区环境质量没有缺陷，也不是说没有改善的必要。问题在于，"场所感""认同感"或"根基感"与物质环境本身的关系远不是诺伯格-舒尔茨所认为的那么简单。

在诺伯格-舒尔茨那里，"生活世界"是"物"的世界。这一点也许最集中地表现于他继《场所精神》后大力倡导的"具象建筑"（figurative architecture）的有关论点里。顾名思义，所谓"具象建筑"以具象为特征。它一反被诺伯格-舒尔茨视为"脱离了现实"的现代建筑的抽象性。那么，诺伯格-舒尔茨自己问道：

> 这里所说的"现实"究竟是何意呢？显然，（现代建筑的）表现的手段也是与现实有关的。色彩、材料、点、线和质地被从我们既有的世界之中抽象出来了，各自以自己的方式反映这个世界。因此，建立在这些元素基础之上的非具象构图也许可以表现某种"真实"的东西。然而，

1 Jan Bengtsson, "Arkitektur och fenomenologi. Om Norberg-Schulz' platsfenomenologi", *Nordisk Arkitekturforskning*, 1994-1, pp.17-33(29).

胡塞尔所谓的“日常生活世界”却丧失了。日常生活世界不是由抽象元素，而是由整体、由具体物组成的。…… 组成世界的是树木和花草，石块和山丘，河流和湖泊，动物和人类，房屋和制品。我们知道、认识并且记得这些物。语言就证实这一点，因为是物则具名。我们称名为实体性的，因为世界的实体就是物。[1]

这样说来，“具象建筑”如何面向事物本身来寻回丧失了的日常生活世界呢？诺伯格-舒尔茨继续道：

基本原型已有名称。我们因而说“塔楼”“边翼”“园厅”“穹顶”“山花”“拱廊”“柱式”“窗户”“大门”“巷子”“街道”“大道”和“广场”。[2]

“具象建筑”说意欲为后现代历史主义建筑语言提供理论基础姑且不论[3]，如此理解“现实”和“生活世界”，诺伯格-舒尔茨建筑理论的不足之处多滥觞于此。

鉴于诺伯格-舒尔茨建筑理论的以上不足，我们应该如何看待建筑的意义问题，如果这一问题本身还具有意义的话呢？西方建筑自意大利文艺复兴之后，随着以伽利略、笛卡尔思想为基础的现代科学的兴起，已逐渐丧失其形而上学的意义，这是原籍墨西哥，现在加拿大麦吉尔（McGill）大学建筑学院主执建筑历史和理论教鞭的佩雷斯-戈麦兹（Alberto Perez-Gomez）在他的《建筑与现代科学的危机》（*Architecture and the Crisis of Modern Science*）一书中经过仔细考察西方建筑中数学几何比例体系性质的变化之后提出的观点。据佩雷斯-戈麦兹说，从 18 世纪法国新古典主义到 20 世纪现代主义，传统的比例体系虽反复延用，却只具工具性而不具这一体系从古希腊到文艺复兴所含的表达宇宙和谐数几关系的形而上学意义。意义的丧失是现代以技术理性和消费文化为主导的“贫瘠时代 ”的表征。然而，意义是人性的基本需求，克服“贫瘠时代”且有待意义的重新探究。诚然，要让食过自然科学之果的现代人再去相信“一切皆为数”或以为宇宙是一

格雷夫斯《走向具象建筑》

1 Christian Norberg-Schulz, “On the Way to Figurative Architecture” in *International Laboratory of Architecture and Urban Design*, Year Book, 1984/85, pp.92-95. 括号中的内容为引者所加。

2 Ibid.

3 关于诺伯格-舒尔茨自己对这个问题的论述，见 Christian Norberg-Schulz, “The Two Faces of Post-Modernism” in *Architectural Design*, 1988-7/8, pp.11-15.

左：海杜克《流浪建筑》　　右：海杜克《白蜡翅膀　金角　石头面纱》英文版

个完美对称的几何形体而津津乐道，就如让失去贞操者重新找回贞操一样不可能。但是按照佩雷斯 - 戈麦兹的理解，源自古希腊的比例体系其本质并不在数几关系本身，它是古希腊人诗性实践（poesis, not merely praxis）的结果。[1] 换言之，意义的重新探究不能通过模仿已成为空洞形式的体系，而必须通过在新的历史条件下重新寻索人之诗性实践方能实现。

据我看，如果佩雷斯 - 戈麦兹这一多少具有海德格尔意味的观点对我们当今的工作有什么启示的话，我们对生活世界的理解就不能见物不见人，而我们对人之诗性实践的理解也不能以过去的建筑环境为准来回避现实。确实，不仅不回避，相反视现实生活世界（the world-as-lived）为意义的第一来源并力反历史主义建筑语言的空洞性，这正是佩雷斯 - 戈麦兹反复强调的主张。尤其值得注意的是，对于佩雷斯 - 戈麦兹来说，通向“生活世界”诗性实践的正是以人的身体性为中介的个体的内在经验。并且，这种诗性实践由于“诗”的自由本质使然也必定是批判性的，不过不是海德格尔意义上的保守性社会批判。佩雷斯 - 戈麦兹继《建筑与现代科学的危机》之后的理论工作，无论是他自己的思想笔记还是对海杜克（John Hejduk）等建筑师作品的分析都说明了这一点。不过，既然这篇文章是为中国读者写的，不妨让我举一个我们自己的例子，那就是李巨川等 1994 年为日本《新建筑》杂志“城市居住”为题的设计竞赛所做的方案。看上去，它既没有从空间上也没有从形式上提供任何具体的建筑方案。有的只是一组方案的作者以一块方砖为伴在城市中的生活照片和这样一段文字：

1 Alberto Perez-Gomez, *Architecture and the Crisis of Modern Science* (Cambridge, Mass. and London: The MIT Press, 1983), p.10.

李巨川 《砖》(1994)

我和一块方砖生活在一起
在城市里。
与此方砖，
我的身体和精神得以栖居。[1]

然而，通过这一看似简单的例子我想说明的是，如果意欲探究建筑的本真意义，我们就必须从当今的现实生活以及人对这一生活的诗性体验和诗性实践入手，而不是自上而下从某个既有的意义或形式入手。如此而来的意义不再是整体和永恒性质上的意义。相反，它可能只来自生活的某一片断，某一偶然事件。它可能只与环境中的一小部分发生关系。它也可能是暂时性的，易变的，含糊不清的。城市再不能作为整体的艺术看待，环境的形成也远非建筑师所能左右。整体破裂之后，意义只在碎片上漂流。在一个变化的世界，意义也随之而变。但是，在这样的基础上来创造空间和形式却还是我们的任务。就此而言，我同意诺伯格 - 舒尔茨在《场所精神》一书最后提出的观点："只有理论是不足以达到我们的目标的，…… 艺术教育比以往任何时候都更为需要。"[2] 这最后一点对于当今的中国来说也许尤为重要。之所以这样说，不仅因为在一个诗情画意的自然景观中需要有艺术的诗性表达，而且在看似乏味的、用诺伯格 - 舒尔茨的观点看甚至是"无意义"的当代城市环境中更需要这种表达。

---

1 见 Alberto Perez-Gomez, *Polyphilo, or The Dark Forest Revisited: An Erotic Epiphany of Architecture* (Cambridge, Mass. and London: The MIT Press, 1992) 或 Alberto Perez-Gomez, "The Renovation of the Body. John Hejduk and the Cultural Relevance of Theoretical Projects" in *AA Files*, No.13. 关于海杜克作品的批判意义，也见 Anthony Vidler, "Vagabond Architecture", *Architectural Uncanny* (Cambridge, Mass. and London: The MIT Press, 1992), pp.207-214.

2 Christian Norberg-Schulz, *Genius Loci*, pp.201-202.

8

强形式与弱形式

——坂本一成访谈[1]

1 最初收录于《反高潮的诗学：坂本一成的建筑》（坂本一成、郭屹民编著），上海：同济大学出版社，2015 年版。

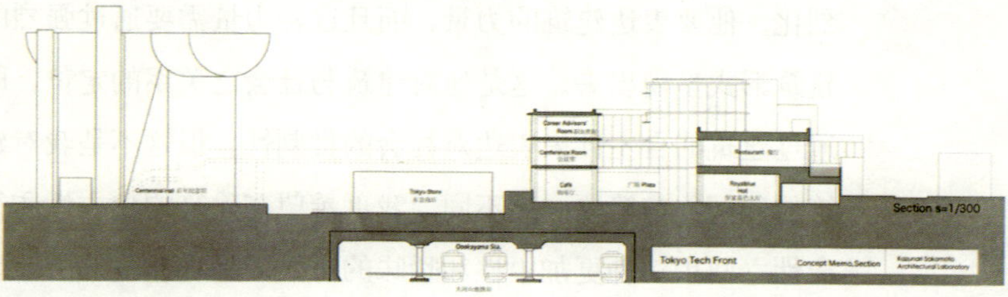

上：东京工业大学百年纪念馆（左）与藏前会馆（右）　　下：东京工业大学藏前会馆剖面图

王骏阳： 坂本老师，非常感谢您能够安排时间接受这次访谈。我为今天的访谈准备了一个提纲，访谈过程中肯定会发生改变，不过我基本想按照这个提纲来进行访谈。

本次访谈主要想围绕坂本老师的 TTF（Tokyo Tech. Front，东京工业大学藏前会馆）提出问题。我们知道在 TTF 之前，您的老师篠原一男先生已经完成东京工业大学百年纪念馆，这是一个形式异常强烈的作品，耸立在东京工业大学的校门口。我在不同的资料上看到您在 TTF 的设计中是想刻意回避百年纪念馆强烈简明的形式，转而追求日常的、更为微妙的、甚至有些晦涩的建筑。当然这也是您许多建筑的特点。但是因为与百年纪念馆并置在一起，所以这一点在 TTF 上就特别明显，是这样的吗？

坂本一成： 日本战后初期，丹下健三的自宅、清家清、池边阳等的住宅都是非常强有力的建筑。此后，日本建筑的重心似乎开始从住宅转向了公共建筑，随之而来的是建筑的社会性成为焦点。即便是住宅，也只有集合住宅被认为是具有价值的建筑，而个人住宅则被排除在外。针对这种现象，篠原一男以住宅的形式作为媒介，对公共建筑一边倒的现象进行批判。我目睹了这一过程，同时也不可避免受到来自篠原老师的影响。其实，在学生时代，我就一直对策划和规划方面很感兴趣，有尝试与策划和规划发生关联的冲动。但是这种冲动很快被篠原一男的空间魅力所消解。通过强烈的形式来呈现空间的魅力也理所当然成为我的目标。

与此同时，我也意识到，与其以对立的方式来呈现建筑的固有性和自主性，倒不如在更大的范畴内或者说在与时代相对应的氛围中来表现更加丰富的建筑内涵。换句话说，尽管篠原一男是我的老师，但我和他毕竟在性格上不同，自然也有着对周遭事物不尽相同的看法，这或许也导致我跟篠原老师不可能走完全相同的道路。从建筑师创作的角度来看，篠原一男创作方式的核心就是将建筑的自主性最大化，或者说强烈化。他要表达建筑的力量，而且这种力量需要通过强烈的建筑形式呈现出来。这是他对建筑与社会之关系的定位，即通过建筑的自主性来表达对社会的批判性。但这不是我对建筑的理解。与篠原一男不同，我更希望在建筑的自主性和社会性之间用一种更加中性和暧昧的方式来呈现建筑的价值判断。即使像 TTF 这样的公共建筑，我也希望把它视为一个具有混杂功能的日常建筑，而不是性质单一的纪念性建筑。就此而言，两者之间的不同确实存在。

王骏阳： 坂本老师讲到 TTF 不是纪念馆，而是一个日常性建筑，有很多功能混杂在一起。但是在功能和形式之间其实没有决定性的关系，而要满足这种混杂功能也应该有其他的形式可能吧。如此说来，我更愿意把 TTF 的形式看作是坂本老师的选择，而不是由功能决定的，更不是时代决定的。

坂本一成： 任何时代都有形式强烈的建筑，也有非常融合的建筑。但我这里想讲的不是建筑形式本身的问题，而是社会对于建筑的定义问题。现代主义是一个封闭边界的时代，事物只在自己定义的范围内发生自主性。而当代正是这种边界被打破的时代，这源自于我们当代人对事物理解的变化。尽管有些事物的形式很强烈，有些事物的形式可能不那么强烈，但是事物本身的定义已经被深刻改变。在这种改变中，我们需要有另一种方式去呈现事物的定义。在我看来，这是我说的时代背景发生根本变化的一个重要前提。如果再补充一点的话，那么也可以把它与我刚才讲的最后一点结合起来进行理解，也就是把它与篠原一男作为建筑师的操作性和我对建筑操作性的不同认识结合起来进行理解。篠原一男是我的老师，我也从他那里学到很多，但从一开始我就感到我们对于建筑的理解是不一样的。也许在中国也会有相同的情况，你可能对于上一代的建筑师的东西接受不了，或者觉得上一代建筑师的想法过于老套。我们在日本同样有这样的感觉。我接受不了篠原一男建筑的强烈形式。篠原希望以超强形式来表达建筑的力量感，但是对于我这一代建筑师来说，这个想法匪夷所思。也许，用这种方式来理解我刚才讲的时代性更加容易些。当然，这样说会使人觉得，我们对于老一代建筑师有些不太尊重，但是我们从一开始确实觉得他们的东西不属于我们这一代人。

王骏阳： 在百年纪念馆的另一侧有一个由安田老师设计的图书馆。按照我的理解，它与篠原一男的建筑很有些相似之处。

坂本一成：（笑）在他们那一代建筑师中，与篠原一男更接近的可能是奥山。当然从大面上看，安田的建筑与篠原确实很接近，而在更为年轻的建筑师当中，与我最有共同之处的应该是塚本由晴。

王骏阳： 在我接受的建筑学教育中有一个设计的基本原则，就是要学会尊重左邻右舍的建筑。比如说因为有百年纪念馆的存在，而且它又是这么重要的一个建筑，所以我们的设计策略应该是积极回应这个建筑，创造一个更好的整体，而不是与它针锋相对，或者我行我素。按照我的观察，TTF建成后，百年纪念馆变得更加孤立了。我的意思是说，如果采取另一种设计策略，从某种意义上对百年纪念馆做出回应，而不是脱离它，甚至反其道而行之，也许对这块城市区域或者说东京工业大学校门口的整体环境来说会更加有利。

坂本一成： 首先我认为王老师的这个设定可能是不成立的，因为百年纪念馆的重要性就在于它的纪念碑意义是无以取代的。如果再有一个类似的建筑在旁边的话，不是消解了它的孤立性，而

安田幸一　东京工业大学图书馆

是造成了校门口的混乱。这种混乱的可能性来源于两个强烈的东西无法调和。这不仅是两者之间的问题，它们与周边的住宅、车站之间的关系也很难调和。就我本人而言，第一我本来也不会做很强烈的房子，这不是我的策略；第二虽然我知道基地边有这么一个重要的房子，但我没有把这个房子变成周围环境的唯一设定。除了百年纪念馆，还有周边的住宅商业街和车站，这些东西是我综合考虑的环境。我做 TTF 的最大愿望就是把这个建筑以一种自然而然的方式融入环境。反过来说，这种策略的一个可能结果也就是让百年纪念馆的纪念意义以更为强烈的方式得到呈现。

王骏阳： 我的意思不是说 TTF 也应该成为一个强烈的形式。我是讲一种可能，TTF 既回应了百年纪念馆，同时又不完全重复它。现在给我的感觉是这两个建筑完全脱离，TTF 完全游离在外，甚至给人一种它不是东工大建筑的感觉。不过坂本老师的意图我多少还是能够领会的，这也是我下面要提的问题。在《自由、解放、中性的建筑空间》这篇文章中，坂本老师把 TTF 归为“通过与现实条件的纠葛产生的构成形式”，这里的“现实条件”不是指百年纪念馆，而是城市，更准确地说是在百年纪念馆和东工大校园旁边的城市。但不可否认的是，百年纪念馆也是非常重要的城市现实，为什么单单忽略它，而去回应周边的其他城市环境？

坂本一成： TTF 与百年纪念馆的联系是一种看不见的联系，而 TTF 与其他建筑的关系是一种同质关系。通过强调与周围其他建筑的连续性，百年纪念馆被从周围建筑中抬得更高了，这就是 TTF 与百年纪念馆的最大联系。与此同时，周围还有很多体量很小的住宅，需要小体量的细分化的形式跟它们形成连续。如何减少 TTF 的高度，以取得与小住宅之间的连续，这是 TTF 思考的。

王骏阳： 除了高度之外，作为一个东工大的公共建筑，TTF 还有什么与小住宅取得连续的方式？

坂本一成： 尽管百年纪念馆体量巨大，它还是一个传统意义上的学校建筑，所有的空间都必须经过一个入口才能进入，所有的空间构成都在内部发生，与外部没有联系。与之不同，小体量的住宅之间形成了半室外的空间，内部与外部之间有着十分暧昧的关系。我想把这种关系延续到通常只考虑内部构成关系

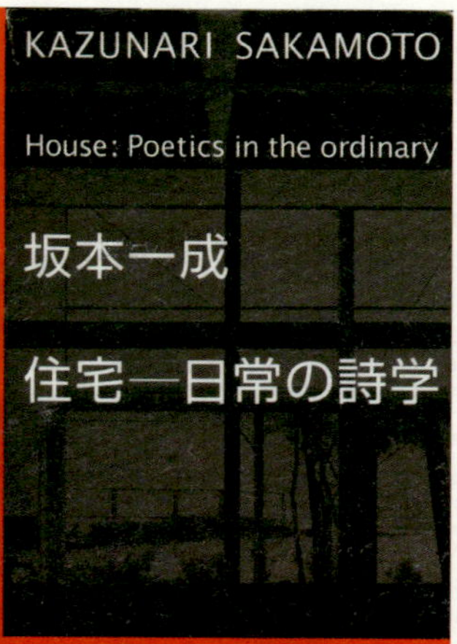

左：从东京工业大学藏前会馆二楼看对面的城市街区　　右：坂本一成《日常的诗学》日、英文双语版

的学校建筑空间之中。TTF 是一个过渡性的建筑，介于学校的固有空间构成关系与住宅、街道的空间构成之间，甚至可以说是介于城市与建筑之间。

王骏阳： 坂本老师的《日常的诗学》上有长岛明夫评论 TTF 的文章。长岛明夫认为，TTF的设计构思是将广场（plaza）、大门（gate）、联系（connection）三个不同的主题通过建筑予以呈现。而您自己在《自由、解放、中性的建筑空间》一文中则说，您的设计意图是从建筑的用途和方位条件出发，通过对广场、大门、连接这三种性格的强化，形成新的城市类型，与既存的建筑类型产生纠葛。我想问的是，这里“既存的建筑类型”指的是什么？是不是刚才坂本老师讲的住宅之间的半外部空间？

坂本一成： 我试图用这三个建筑元素——广场、大门、连接形成一个由交通和流线组成的 network（关系网）。这三个元素从来没有被组合在一起过，当它们第一次被组合在一起的时候，就会引发新的结构，也就是在与周围的街道和住宅这些我们司空见惯的功能之间的关系方面形成不一样的结构。这种新的结构是我在这个建筑之中试图寻找的。这里所谓的“新”是针对特定的项目才能够出现的一种结构。每一个建筑都有不同的形式，这些不同形式的出现是因为它有不同的结构关系。在我看来，建筑师需要重视建筑物的类型，尽管类型也很容易把建筑学变成修辞性的工作。要在公共建筑中将类型完全摒弃掉是非常困难的，而在日本的学校建筑中就更难。

王骏阳： 那么新的城市建筑类型是什么？在 TTF 中产生了什么新的建筑类型？

左：东京工业大学藏前会馆入口　　中：施工中的藏前会馆讲演厅
右：藏前会馆讲演厅与在二层穿过其中的露天步行走道

坂本一成： 就是我刚才讲的，三种从来没碰到过一起的组合类型，以及由此形成的新的结构关系。

王骏阳： 如果以大门为例，为什么要在上面做一个百页格珊的屋顶？长岛明夫先生在他的文章中说，这是因为TTF缺少某些识别性，要通过百页的屋顶强化体量之间的统一感，是这样的吗？

坂本一成： 大岛的文章具体怎么写的我已经不记得了，但是这个百页屋顶绝对不是为“门”而做的。首先TTF不仅仅是个学校校门，如果仅仅是学校校门，也许它会有另外一种方式来呈现。通过这个百页屋顶，我希望将下面分置的体块暧昧化，使形象从眼前后退，变得更加不清晰，这是百页屋顶存在的一个重要意图。

王骏阳： 换句话说，不是为统一的大门形象，而是要使形象更加暧昧化。

坂本一成： 对，绝不是要做一个“框”把这个门标志出来，没有这样的意图。

王骏阳： 坂本老师在《自由、解放、中性的建筑空间》中还提到“另一种现实和日常”，指出它“不仅仅是通过与现实和日常对抗的新的概念形式来实现的，而且是通过构成现实和日常的建筑形式自身产生变形、错位而成立。”如果以TTF为例，这种变形和错位究竟体现在什么地方？它们又是怎么发生的？

坂本一成： 还是以门作为例子，它上面有个百页屋顶，为比例关系的变化提供了尺度。如果你去看过的话应该注意到地面是斜的，当人在百页屋顶下方通过的时候，这个斜度形成地面与屋顶之间的尺度变化，同时感觉这是学校的一个出入口，但是具

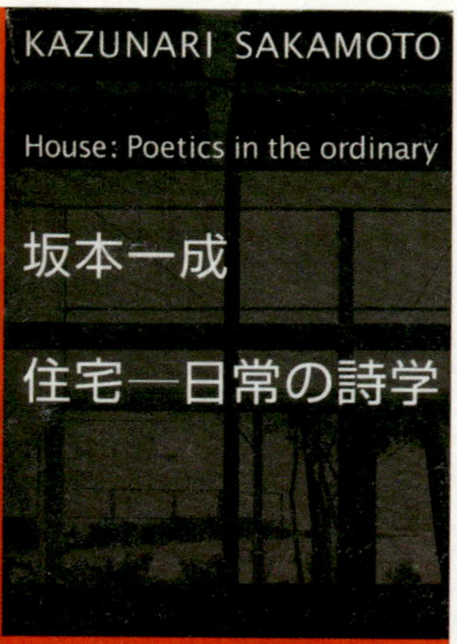

左：从东京工业大学藏前会馆二楼看对面的城市街区　　右：坂本一成《日常的诗学》日、英文双语版

的学校建筑空间之中。TTF 是一个过渡性的建筑，介于学校的固有空间构成关系与住宅、街道的空间构成之间，甚至可以说是介于城市与建筑之间。

王骏阳： 坂本老师的《日常的诗学》上有长岛明夫评论 TTF 的文章。长岛明夫认为，TTF 的设计构思是将广场（plaza）、大门（gate）、联系（connection）三个不同的主题通过建筑予以呈现。而您自己在《自由、解放、中性的建筑空间》一文中则说，您的设计意图是从建筑的用途和方位条件出发，通过对广场、大门、连接这三种性格的强化，形成新的城市类型，与既存的建筑类型产生纠葛。我想问的是，这里“既存的建筑类型”指的是什么？是不是刚才坂本老师讲的住宅之间的半外部空间？

坂本一成： 我试图用这三个建筑元素——广场、大门、连接形成一个由交通和流线组成的 network（关系网）。这三个元素从来没有被组合在一起过，当它们第一次被组合在一起的时候，就会引发新的结构，也就是在与周围的街道和住宅这些我们司空见惯的功能之间的关系方面形成不一样的结构。这种新的结构是我在这个建筑之中试图寻找的。这里所谓的“新”是针对特定的项目才能够出现的一种结构。每一个建筑都有不同的形式，这些不同形式的出现是因为它有不同的结构关系。在我看来，建筑师需要重视建筑物的类型，尽管类型也很容易把建筑学变成修辞性的工作。要在公共建筑中将类型完全摒弃掉是非常困难的，而在日本的学校建筑中就更难。

王骏阳： 那么新的城市建筑类型是什么？在 TTF 中产生了什么新的建筑类型？

左：东京工业大学藏前会馆入口　　中：施工中的藏前会馆讲演厅
右：藏前会馆讲演厅与在二层穿过其中的露天步行走道

坂本一成： 就是我刚才讲的，三种从来没碰到过一起的组合类型，以及由此形成的新的结构关系。

王骏阳： 如果以大门为例，为什么要在上面做一个百页格珊的屋顶？长岛明夫先生在他的文章中说，这是因为TTF缺少某些识别性，要通过百页的屋顶强化体量之间的统一感，是这样的吗？

坂本一成： 大岛的文章具体怎么写的我已经不记得了，但是这个百页屋顶绝对不是为“门”而做的。首先TTF不仅仅是个学校校门，如果仅仅是学校校门，也许它会有另外一种方式来呈现。通过这个百页屋顶，我希望将下面分置的体块暧昧化，使形象从眼前后退，变得更加不清晰，这是百页屋顶存在的一个重要意图。

王骏阳： 换句话说，不是为统一的大门形象，而是要使形象更加暧昧化。

坂本一成： 对，绝不是要做一个“框”把这个门标志出来，没有这样的意图。

王骏阳： 坂本老师在《自由、解放、中性的建筑空间》中还提到“另一种现实和日常”，指出它“不仅仅是通过与现实和日常对抗的新的概念形式来实现的，而且是通过构成现实和日常的建筑形式自身产生变形、错位而成立。”如果以TTF为例，这种变形和错位究竟体现在什么地方？它们又是怎么发生的？

坂本一成： 还是以门作为例子，它上面有个百页屋顶，为比例关系的变化提供了尺度。如果你去看过的话应该注意到地面是斜的，当人在百页屋顶下方通过的时候，这个斜度形成地面与屋顶之间的尺度变化，同时感觉这是学校的一个出入口，但是具体在哪个位置是确切的“门”却不能辨识。这就是我一直想说的，一种你从来没有看到过的但是确实成立的现实存在的可能性。通常，这种可能性被我们的一些固化的东西排斥掉了，而我希望通过不同事物的叠合把它重新呈现出来。这种呈现既不强烈也不清晰，但它是存在的。

王骏阳： 确实，TTF 在很多方面都很暧昧，甚至平淡无奇，这肯定是坂本老师刻意为之。但在整个建筑中间有一个元素给我非常深刻的印象，这就是从建筑最外侧的大台阶穿过建筑和报告厅通往旧馆的一个斜向通道。这个元素非常强烈，似乎与您说的暧昧性，也就是做一个东西的同时还要把它化解的策略不是很吻合，这一点怎么解释？

坂本一成： 这条斜线从图形上来看的确非常强烈。但是图形上的强烈并不等于实际的空间体验，因为形式的强烈与否很大程度上取决于人观看它的方式。当一个形式正对着人出现的时候，它的强烈性会凸显，而我试图减少正面性。因为没有一个强烈的面来对应它，所以这个斜廊只是从你身边过去而已。但是它的确存在，而且它的存在也很重要，所以需要有一种方式在呈现其存在的同时，又不让它从人视线的正中央穿越。我试图用这种方式将城市的街道和空间的关系呈现出来，也就是让城市的街道能够在建筑中以一种“存在”的方式而不是“视觉”的方式呈现。这是我想做的。

另外，这条线看上去有点强的原因是这条路现在断头了。非常不幸的是，原来要做这个桥的校长在造桥之前换届了，而新的校长不喜欢这个桥，结果不了了之。如果这条道路能够穿越东急超市和道路进入学校，真的成为学校和社会的一条通道的话，那么这条路在我看来就会有助于将都市的自然导入封闭的校园内部。

我刚才无意中讲到了都市的自然，这个词我其实没有经过深思熟虑。我想表达的是，无论自然形成的城市，还是经过规划的、有历史底蕴的城市，当你在城市当中走的时候所感受的城市的自由，以及无拘无束的状态，这些恰恰是建筑常常缺少的，所以才希望这种自然也出现在建筑之中。怎样把城市的自然最大限度地与建筑结合在一起是我刚才想表达的意思。

我非常不喜欢机械化、图案化的东西，这些东西与我所谓的都市的自然是完全不同的概念。

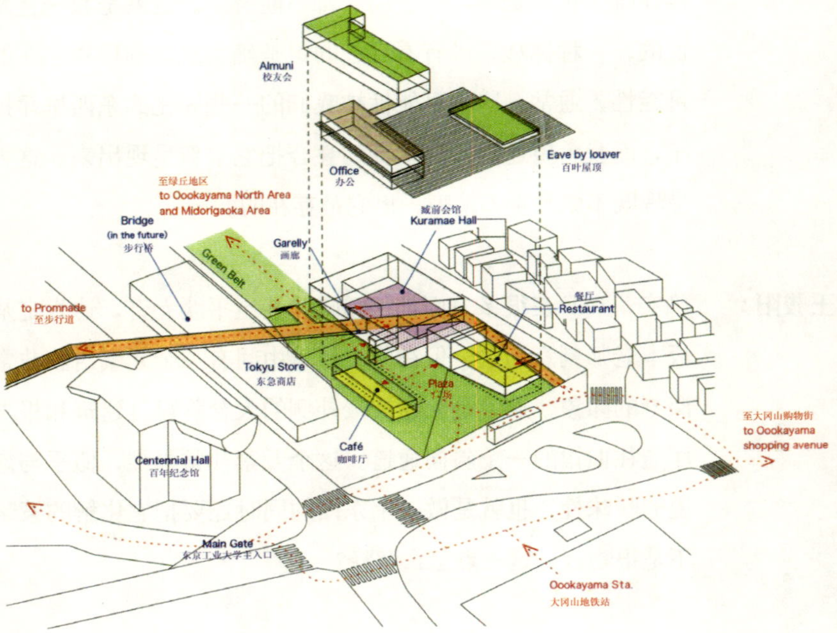

东京工业大学藏前会馆图解

王骏阳： 回到《自由、解放、中性的建筑空间》这篇文章，其中坂本老师特别讲到自由的追求。看起来这个自由是指自由的建筑空间，而不是自由的建筑形式。坂本老师怎么看两者的区别？

坂本一成： 我认为最高的自由是一种身心的自由，没有精神上的枷锁，把一切限制自己的精神压抑全部解脱之后的随心所欲的自由。当然，这种自由在社会当中不应以他人的不自由作为代价，也就是说自由应是均衡的自由。对于使用者来说，这种自由的确很重要，这意味着（建筑）不应对你产生强烈的视觉干扰。我想做的自由是将所有的限制统统取消以后的自由。某种程度上说，这才是设计师的自由。

王骏阳： 我们处在电脑和数字化的时代，自由的建筑形式似乎越来越多，但是在坂本老师看来这些也许只是“过分自由”。您追求的是“更为温和的自由”或者说“中性”的自由。应该如何理解这里的“温和”“中性”呢？应该怎么来把握它？是与“日常”的关系吗？

坂本一成： 以刚才说的那个斜线为例，如果设计师为了自由把它变成曲线，这样做看似有了一个自由形态，但是这样一个曲线的空间对人来说其实非常不自由，因为它的控制性太强，迫使人按曲线方式行走，这跟纯粹通过性的、看起来什么样都没有的平淡如水的道路所具有的自由度不可同日而语。这也是我

刚才讲的自由不能以牺牲他人的自由为代价的一个方面，是我们在谈论自由时需要把握的。所谓过分自由，就是你的自由是建立在他人的不自由的基础上。这是“中性”“温和”的自由应该避免的。

在计算机时代，什么样的形式计算机都可以画出来。但是，如果说这些画出来的东西在白纸上呈现出一种自由创作，那么这些自由的创作能否真正呈现自由是非常值得怀疑的。真正的自由与过去、习惯以及我们的身体的经验有关。在我看来，摆脱和它们的关联去创造新的自由是不可能的。自由的呈现一定要回到现实当中来，回到生活当中来，只有这样才能把自由还原。

王骏阳: 除了TTF之外，我还参观过坂本老师设计的几个住宅项目，如水无濑、House SA，还有代田的町家等。我感到使用者在这些住宅中都需要小心翼翼，而且是几十年如一日，不能对建筑空间、内部格局甚至家具做太多改变。在这种意义上来看，我觉得这些住宅似乎剥夺了人的自由，坂本老师对这一点怎么看？

坂本一成: （笑）这个问题提得非常好。代田的町家已经有三十多年没有动过了，即便有也是很小的动作。其他房子也差不多。这的确会造成一种印象，就是人必须适应它们，而不是它们来适应人。但是我在做住宅设计的时候，一般不会太关注住宅方面的细节，只对大致的家庭构成以及生活方式做一个宽泛的指标性的安排。剩下的具体内容比如房间的定义或者流线等基本上是一个自由的状态。反过来说，它的宽泛性和容忍度其实是很高的。相比之下，大多数房子由于分隔太细，所以会经常变动，比如换了业主就会把它变掉，而我的房子即使在父母去世之后由子女来使用，他们也不愿意对它有太多的改变，因为即使保持功能流线不变也是可以接受的。从这个角度来说，我设计的房子是一种非常自由的房子，它们不需要做太多的改变也可以供下一代使用。王老师可能与这个房子接触的时间比较短，而委托我设计这些房子的业主是真正住在其中的，他们并不愿意进行太大的改变。

关于代田的町家，前两天刚出版的《新建筑》上有一个报道，很多作者谈了参观这个建筑之后的感想。其中一位是著名建筑评论家上田石，他文章的观点与王老师非常相似。他认为这个建筑物是个非常“强”的建筑物，但是这种“强”是从

坂本一成
左：水无濑　　右：House SA

建筑学角度看的。对于居住者来说，即便想对它改动也无从入手，因为它是一个整体，无法改变它的一个部分来改变建筑师对于建筑的空间定义和性格的控制。但是另一方面，你却感觉不到它在非建筑学层面上对住在其中的使用者的生活方式的控制，这正是这个建筑物非常奇怪的两面性，即在建筑学层面的强烈性和在空间自由度上的宽泛性。这也许就是王老师明显感受到的一点，所以我说您提的问题非常好。

这里讲的建筑的“强”和百年纪念馆的“强”是不一样的，百年馆的强是单向的，它通过建筑学本身的强烈性实现对内部使用者的强烈控制。这不是我想要的。在我的建筑中，“强”是双向的，当使用者不具有建筑学概念的时候，他不会感到这个房子强大的存在，但是如果你掌握建筑学知识，那么你会捕捉到建筑学的自主内容。这是建筑学的另外一种“强”，也是它不同于篠原的地方。

王骏阳： 如果把百年纪念馆视为一种“强形式”，那么是否可以延伸意大利哲学家瓦提莫（Gianni Vattimo）的“弱思维（weak thought）”概念，把坂本老师的建筑称为“弱形式（weak form）”？

坂本一成： 首先非常抱歉我讲的话可能非常抽象，给王老师造成一定的误解。我认为对于建筑学来说，建筑形式非常重要。建筑不可能通过自然而然的方式来形成，这种自然而然的东西不是建筑。

我想强调的是，如果没有形式，建筑就不是建筑。但是我也一直认为，形式只是一种手段，是在一个更加宽泛的社会层

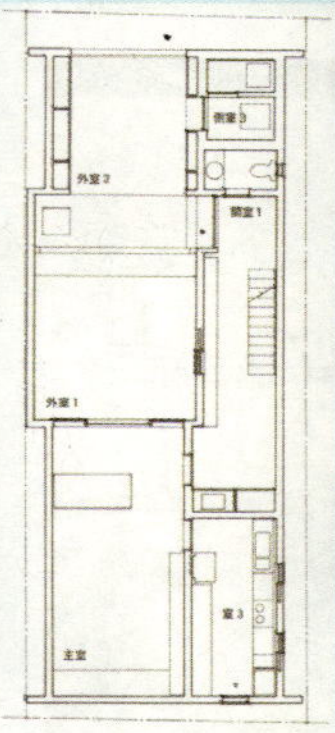

坂本一成：代田的町家

面使建筑学成立的手段。当这种形式不是手段而是目的的时候，它就变成形式主义。这不是我想要的东西。我希望形式是在和各种内外关系的综合之中成立的东西。换言之，当形式作为一种手段的时候是好的，但是如果停留在表层，或者为形式而形式，这不是我的目标。篠原一男的形式是目的，而我的形式是手段。

王骏阳： 是否可以说“弱形式”就是让形式成为手段？

坂本一成： 对的。

王骏阳： 您目前在中国同时有三个项目在做。这些项目与您之前在日本完成的项目肯定有很多不同。我感兴趣的是这几个中国项目有没有在一定程度上改变您之前对建筑的思考，还是仅仅因为中国的条件不同，只是寻求不同的解决办法而已？

坂本一成： 在中国一下拿了三个项目，而且是非常不同的项目，要说跟日本的项目完全不一样固然未尝不可；但是若要说跟日本的项目很相似，也不是完全没有道理，关键是从哪个角度来说。总而言之，这三个项目给我之前在日本没有的经验。宁波项目是一个一千多平方米的豪宅，这种住宅之前在日本有，但现在已经很少见。这一千多平方米放在日本要如何消化掉？在日本，住宅是给人住的，而中国的豪宅有很多地方是给人看的，比如艺术收藏。从这个角度来说，这个住宅是一种全新的 building type（建筑类型），即一种把公共性导入住宅内部的类型。这给我非常大的新鲜感。第二个项目是上海当

坂本一成：常州工学院宾馆（设计中）

代艺术博物馆的图书室，它试图营造一种“书山”的感觉，但是每一本都是人伸手可及。这样的项目在日本不是不能设想，但是恐怕没有这样的业主。日本有很多图书馆，但是这些图书馆的书大多用手拿不到。这不是我想要的图书馆。我想要的是你可以随心所欲拿到书，同时有一种你被书山包围的感觉。对我来说，这又是一种全新的 building type。第三个项目是常州工学院的 Guest House（宾馆）。首先，大学拥有一个酒店宾馆，这在日本很少见；其次，作为酒店，它又是学校面向社会的一个窗口，拥有双重功能，或者说双重性格；第三，它会容纳不同的人，有留学生、专家和社会人士。这些 building types 是我之前从未碰到过的。在我看来，新的 building type 必将导致新的形式和建筑的可能性，这是我愿意接受的挑战。它们给我强烈的新鲜感和尝试的动力，也能满足我自己挑战建筑的愿望。我在中国做这些项目感到非常愉快。现在中国什么样的东西都有，什么都同时存在，不仅有日本的建筑师，世界上几乎所有的建筑师都来了。但是与很多建筑师不同，我对形式实验游戏没有兴趣。

坂本一成： 能够面对只有在中国才能遇到的 program（建筑的内容纲要），能够用我的方法实现空间结构的新的可能性，我感到很幸运。不知这样能否很好回答王老师的问题。（笑）

王骏阳： 非常期待坂本老师能够完成这些项目，也非常感谢坂本老师能够接受我的访谈。

9

从皮拉内西“考古学面具下的重构”到库哈斯“偏执妄想的批判之旅”

——关于《小、中、大、特大》的读书笔记[1]

1 最初以“Piranesi, Koolhaas and the Subversion of the Concept of Place”发表于《北欧建筑研究》（*Nordic Journal of Architectural Research*） 1997 年第 2 期，收录本文集的中文译文与原文有较大差异。

在20世纪后半叶的建筑师之中，或者更准确地说是关于建筑和城市的思想者之中，雷姆·库哈斯（Rem Koolhaas）当属最具有挑战性而又最发人深省者之列。具有挑战性的是他对建筑思想界既有思维方式的质疑，无论这些思维方式是以“历史价值”或“文脉主义”的常用语言、还是以海德格尔哲学的修辞面貌出现。在人们以一种怀旧的心态大谈特征和可识别性失落之时，库哈斯直面那些形成20世纪都市文明的众多力量。发人深省的是他在接受20世纪城市发展之现实的同时努力思考和挖掘事物新维度的策略。他在庸俗中看到崇高，在恐惧中看到希望，在精神分裂式的城市状态中看到理性。他在弥漫的都市原野中寻找隐藏的规律，在当代变化无常的世界中探索潜在的可能性。

OMA/库哈斯/布鲁斯·冒
《小、中、大、特大》英文版(1995)

由库哈斯、大都会建筑事务所（The Office for Metropolitan Architecture，简称OMA）和平面设计师布鲁斯·冒（Bruce Mau）联合署名的新书《小、中、大、特大》（*S, M, L, XL*）就是他上述思想工作的一次集中展现。该书集散文、小说、日记、旅行日志、方案、图纸、模型、照片、漫画和报纸广告为一体，不仅反映出库哈斯建筑实践丰富多样的形式，也打上了他早年记者和编剧生涯的深刻烙印。书名“小、中、大、特大”借用服装尺码的标注方法，似乎表明作者是根据建筑或者设计项目的规模组织书的内容，但是在建筑作为思想之载体的意义上来说，《小、中、大、特大》也意味着思想的范围，或者更确切地说，被思考对象的量级。

但是除了S、M、L、XL的划分之外，该书似乎并没有一部著作通常具有的章节结构和版面安排。异乎寻常的厚度（1345页）和重量（1.5公斤）、海量的内容、文字、草图、图纸、模型、照片的交错出现、篇幅上占大多数但刻意不讲究印刷精度的图片、字母尺寸在大小之间跳跃式的变化等，反倒是它给人最为强烈的第一印象。细读起来，书中的文字有些是用来说明建筑或者设计项目的，但也有相当一部分属于自成一体，与建筑或设计项目没有直接联系，更不是项目的文字说明。更值得一提的是，作者刻意在书中不连贯地穿插安排了一个从A到Z的词条解释，它时有时无，看似不起眼，但却是该书多重复合结构中不可或缺又完全可以独立阅读的组成部分。显然，书的作者和设计者的意图在于尽

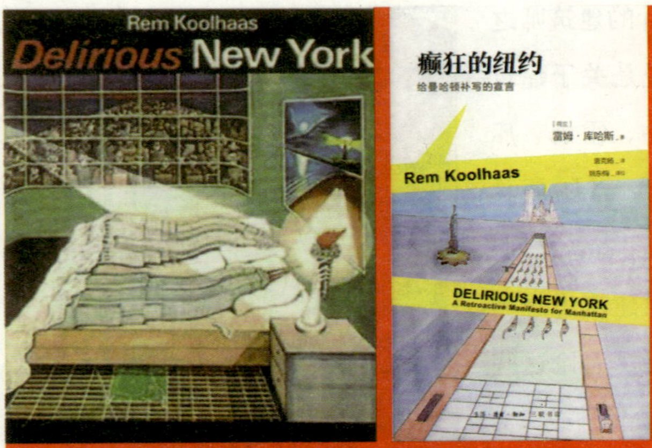

库哈斯 左：《癫狂的纽约》英文版（1978） 右：《癫狂的纽约》中文版（2016）

可能多地颠覆和削弱传统的线性阅读方式，凸显非线性甚至混乱的主题。用库哈斯在“引言”（Introduction）中的话来说，“建筑本质上就是一种混乱的历险”。（Architecture is by definition a chaotic adventure.）[1]

在此，人们不禁想起尼采（Friedrich Nietzsche）、德勒兹（Gilles Deleuze）和瓜塔里（Felix Guattari）等现代思想家践行的“游牧思维”（nomad thought）。它试图在光滑开放的流动空间中运作，可以在任何一点出现，也可以随时滑向另外任何一点。它努力摆脱桎梏，争取自由，无论这种桎梏是意识形态的、制度的还是专业层面的。游牧思维源自游牧者狂野心灵的渴望和想象，是对经验和思想极限的超越，它将自己置于内部世界和外部世界永不停息的征途之上。库哈斯的内在征途与他的外部征途并行不悖，《小、中、大、特大》既是他过去数十年世界之旅的见证，也是他对政治、经济和全球化的影响之下的 20 世纪建筑和城市环境进行思考的结果。

在他还没有任何真正的建成项目之时，库哈斯就曾因一部《癫狂的纽约》（*Delirious New York*）一举成名；现在这部《小、中、大、特大》同样由纽约的蒙纳切利出版社（The Monacelli Press）出版，更不要说 OMA 最初成立于纽约这个基本事实。作为一个荷兰人，库哈斯似乎与纽约有着某种不解之缘——《癫狂的纽约》甚至用法国蚀版画家约兰（Jollain）1672 年的《新阿姆斯特丹鸟瞰》来比拟曼哈顿。然而，曾经是 OMA 成员、后成为柏林工业大学建筑历史理论教授的弗里茨 · 纽迈耶（Fritz Neumeyer）指出，库哈斯世界之旅的出发点——因而也是 OMA 的出发点——“不是人们可能会认为的纽约，而是柏林”。[2]

那是在伦敦建筑联盟建筑学院（AA）就学期间，受“直觉和 60

1 OMA, Rem Koolhaas and Bruce Mau, *S, M, L, XL* (New York: The Monacelli Press, 1995), p.xix.

2 Fritz Neumeyer, "OMA's Berlin: The Polemic Island in the City", *Assemblage* 11 (April 1990), p.38.

年代末那种与日俱增的天真、内心的烦恼以及率直的记者兴趣”的驱使，[1]库哈斯为自己设定了一个以“柏林墙作为建筑”（Berlin Wall as Architecture）的毕业设计主题。一年后的1972年，库哈斯与当时已是AA老师的伊利亚·曾吉利斯（Elia Zenghelis）合作，在意大利《卡萨贝拉》（*Casabella*）杂志举办的“一个有意义的城市环境”（A City with a Significant Environment）的设计竞赛中提交“大逃亡——自愿成为建筑的囚犯”（Exodus, or the Voluntary Prisoners of Architecture）的参赛作品。在这个作品中，“柏林墙作为建筑”的思路再度出现，不过是以一个发生在伦敦的虚构故事的形式出现：“从前，一个城市被分为两半。一半是好的，另一半是坏的……”[2]

这个所谓“好的一半”是一个贯穿伦敦市中心狭长地块的“人造天堂”，边界上的两堵大墙将其从现有城市隔离出来。与库哈斯早在进入AA学习建筑之前就已经敬仰的俄国构成主义者伊万·列奥尼多夫（Ivan Leonidov）1930年的莫斯科文化宫（Palace of Culture）设计竞赛方案相类似，隔离区域内部被划为一连串主题各不相同的正方形地块——自留园地、社区设施，等等。这个“好的一半”犹如疲态尽现的大都市理想中注入的一针强心剂，受它的刺激，已经陷入麻木的伦敦居民蜂拥而至，最终形成一个自愿成为其囚犯的大逃亡。

这并非一个一般意义上的虚构故事。就其宣称的一个足以使人们趋之若鹜的“美好”天堂而言，它更像是一个乌托邦主义的产物。但是，正如建筑史学家迪米特里·波菲里奥斯（Demetrios Porphyrios）在1970年代初期已经指出的，“大逃亡”既不是历史上第一个城市乌托邦，也

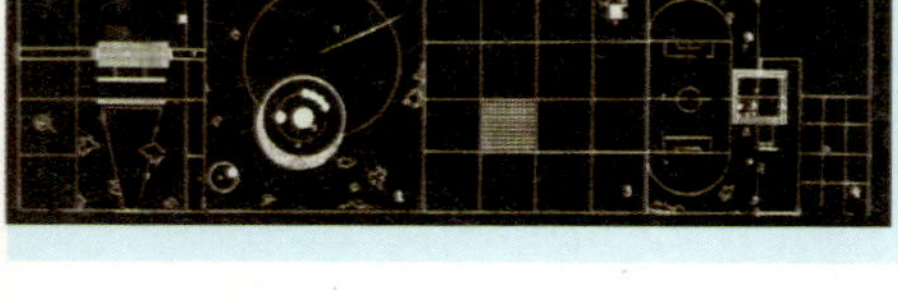

左：列奥尼多夫　莫斯科文化宫设计方案（1930）
右：库哈斯 & 曾吉利斯
《大逃亡——自愿成为建筑的囚犯》

1 *S, M, L, XL*，*p.216*
2 Ibid, p. 5 .

与之前的乌托邦截然不同。它无意为治愈或救赎既有城市的弊端提供解决方案。[1] 相反，在看似为城市机体插入具有完全理想化的作为现实之替代产品的“社会聚化器”（social condenser）[2] 的同时，它揭示了这样一个无情的事实：任何社会完美都有其阴暗面，建筑可以作为监禁的工具。更准确地说，它呈现的是一种“通过集体设施充分满足个人欲望的肆无忌惮的设计学”（hedonistic science of designing collective facilities that fully accommodate individual desires）[3]，由此引发的“建筑学冲突”（architectural warfare）宛如乔治·奥威尔（George Orwell）的小说《一九八四》中的场景：炼狱般的接待区，极权监视下的自留园地……

事实上，除了展现了一幅幅劫难中的文明图景——或者说一部“图像启示录”（apocalipsis cum figuris）之外，“大逃亡”描绘的理想化都市原型（“好的一半”）没有提供任何乌托邦的允诺。与传统意义上的乌托邦美好构想迥然不同，它更像是“龌龊现实主义”（dirty realism）[4] 的建筑范本，一个与柏林墙这个遍体鳞伤的 20 世纪都市的极端代表进行剧烈心理碰撞的结果。

因此，从他的建筑生涯之始，库哈斯就已经抛弃了“优雅建筑学”（polite architecture）的智识基础。如果说库哈斯思想中确实有某种乌托邦色彩的话，那么它无疑是一种反乌托邦的乌托邦——或者，借用已故意大利评论家和历史学家曼弗雷多·塔夫里（Manfredo Tafuri）的术语

库哈斯 & 曾吉利斯《大逃亡》中的场景　左：到达与接待　右：自留地

1　Demetrios Porphyrios, “Pandoras Box,” *Architectural Design* 47，1977-5, p.357.

2　“社会聚化器”是俄国构成主义的建筑理论，由 M. 金兹堡于 1928 年提出，其中心思想是建筑具有对社会行为施加影响的能力以及建筑中的集体主义。《小、中、大、特大》对 condenser 的词条解释是：“纽约下城健身俱乐部的摩天大楼就是一个构成主义的社会聚化器，一部引发和强化理想的人类互动形式的机器。”

3　同上。

4　“龌龊现实主义”，美国 20 世纪文学中的批评术语之一，由英国文学杂志《格兰达》（*Granta*）的主编比尔 · 巴福德（Bill Buford）与 1983 年提出，以概括此前一、二十年在北美文学中出现的一种新现象，其特点是简单甚至单调的语言呈现日常生活中更为平凡或者丑恶的方面。代表人物有雷蒙 · 卡佛（Raymond Carver）等。

来说，它是一种"否定的乌托邦"(negative utopia)[1]。在这里，想象与湮灭、总体与虚无、乌托邦和反乌托邦之间的界限荡然无存。

纵观建筑历史，以如此方式思考建筑学的案例可谓凤毛麟角。一个可比的人物是 18 世纪意大利建筑师和蚀刻画家皮拉内西（Giovan Battista Piranesi）。他的《古罗马战神原》组图中的城市平面是《小、中、大、特大》一千三百余页中为数不多的几个 20 世纪之前的建筑图像之一。

塔夫里《球与迷宫——从皮拉内西到 1970 年代的先锋派与建筑》英文版（1987）

一 皮拉内西与"古罗马战神原"的"考古学重构"

据我所知，塔夫里是将皮拉内西的作品与 20 世纪现代建筑（而不是 18 和 19 世纪浪漫主义）联系起来进行研究的第一人。他的《球体与迷宫》（*The Sphere and the Labyrinth*）进一步发展了之前《走向建筑的意识形态批判》（"Toward a Critique of Architectural Ideology"）[2] 和《建筑与乌托邦：设计与资本主义发展》（*Architecture and Utipia: Design and Capitalist Development*）已经开始的研究工作，以"'恶作剧的建筑师'：G.B. 皮拉内西、异托邦与远游"（"'The Wicked Architect': G.B. Piranesi, Heterotopia, and the Voyage"）作为该书的开篇之章。诚如塔夫里自己所言，从皮拉内西的作品开始对先锋派与建筑学的关系进行研究，这无疑具有某种挑衅性。但是，作为现代主义先锋派电影的先驱之一的谢尔盖 · 爱森斯坦（Sergei Eisenstein）曾经在写作中反复引用皮拉内西，这在一定程度上为塔夫里的先锋派历史分析和它的"历史性"（historicity）研究提供了线索乃至合法性。[3] 在塔夫里那里，正是通过皮拉内西《监狱》组画（*Le Carceri*）虚构的戏剧效果，想象的主题以及它全部的意识形态意义得以进入现代建筑学的历史视野。另一方面，正如爱森斯坦电影的蒙太奇（Montage）手法所表明的，历史先锋派又将《监狱》组画的空间和时间观念推向新的发展。

---

1 Manfredo Tafuri, *The Sphere and the Labyrinth: Avant-Gardes and Architecture from Piranesi to the 1970s*, trans. Pellegrino d'Acierno and Robert Connolly (Cambridge, Massachusetts and London: The MIT Press, 1987), p.29.

2 Manfredo Tafuri, "Toward a Critique of Architectural Ideology", *Architecture Theory since 1968*, ed. Michael Hays (Cambridge, Massachusetts and London: The MIT Press, 2000), pp.6-35.

3 Tafuri, *The Sphere and the Labyrinth*, p.55.

皮拉内西《监狱》组画（1761）

这是一个双重的历史交汇。作为这个交汇点的载体，塔夫里在对《监狱》组画的介绍中指出，“必须一开始就明确，除了可能引发的情感反应之外，所有这种破碎、扭曲、重复、无序恰恰是以视觉传达工具为手段，对场所概念（the concept of place）进行的系统批判。”[1] 乍看起来，塔夫里在接下来的研究中似乎并没有对场所概念的“系统批判”作进一步阐述，不过，考虑到过去数十年有关场所概念的讨论，我们不难看出塔夫里的提示应该具有的意义。这是因为，如果场所概念本质上与“中心”相联系，那么《监狱》组画中变幻莫测的空间无疑是“对‘中心’概念的系统批判”[2]；如果场所概念需要以可识别的建筑语言为前提，那么《监狱》组画中热情交织的结构以及对透视法的破坏和摆脱形式的自由无疑都是“对作为‘作用于世界的一种方式’的语言的明确批判”[3]；如果场所概念的合法性在于主体无法摆脱对历史和自然的依赖，那么，在《监狱》组画的“消极乌托邦”之中，“历史与自然已经与主体分道扬镳，这不是打开了一个新的价值世界，而是将这种剧烈的断裂作为唯一可能的价值”。[4]

塔夫里对《监狱》组画的上述分析基本上都是与皮拉内西的另一作品《古罗马战神原》组图（*Il Campo Marzio dell'Antica Roma*）联系在一起的。后者发表于1761—1762年间，也就是《监狱随想》组画（*Invenzioni*

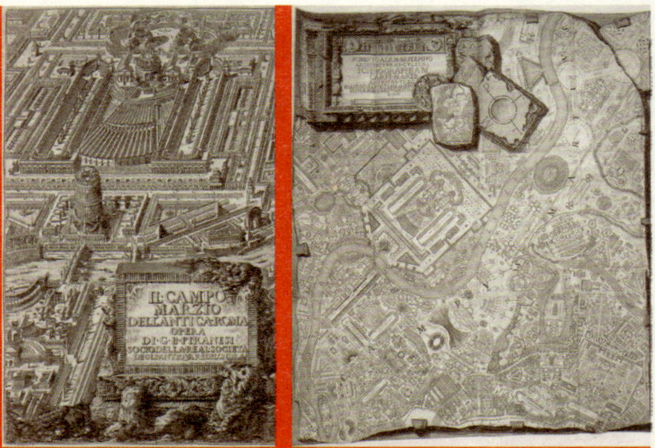

皮拉内西《古罗马战神原》组画之一、二（1761—1762）

1 Tafuri, *The Sphere and the Labyrinth*，p.27.

2 Ibid.

3 Ibid, p.34.

4 Ibid, p.38.

Capricciose di Carceri）的修订版完成之后不久。然而，两者之间的差异也不容忽视。首先，如果说《监狱》组画勾画的都是看似子虚乌有的场景，那么《古罗马战神原》组图则将我们直接带到一个特定城市——罗马。从《建筑与透视第一部分》（*Prima Parte di Architettura e Prospettive*）开始，继而是《建筑、透视、怪物、古迹》（*Opere Varie di Architettura, Prospettivi, Groteschi, Antichità*）和四卷本《罗马古迹》（*Le Antichità Romane*），皮拉内西一直致力于对罗马建筑和工程成就的考古学调查。其次，与《监狱》组画所呈现的疯狂——或者用爱森斯坦的话说是"'癫狂'的构建方法"（the "ecstatic" method of constructing）[1]不同，《古罗马战神原》组图的考古学方法似乎带有某种理性和"科学研究"的基础。有学者甚至用"基于理性的学术方法，也合乎启蒙运动的精神及其对历史变化现象的关注"[2]来形容它。皮拉内西本人在给苏格兰设计师罗伯特·亚当（Robert Adam）的信中也强调，他不仅对卡比多（Capitol）古城区的平面进行了仔细研究，而且还将这一研究与当时最优秀的历史学家的知识相结合。他因此断言"人们不会将我所做的一切视为仅仅没有任何证据的奇思异想。"[3]就此而言，人们似乎不能把《古罗马战神原》组图等同于《监狱》组画那样一种纯粹想象的产物。

然而这正是《古罗马战神原》组图的有趣之处。一方面，正如斯坦·艾伦（Stan Allen）曾经注意到的，一个看似致力于对罗马古城进行"考古学重构"（archaeological reconstruction）的研究却明显忽视了这个城市的历史中心，尽管那里聚集了大量的历史遗迹。[4]取而代之的是，皮拉内西将他的注意力转向"战神原"地区，而该地区在历史上一直处在以公元前 4 世纪就已经存在的以塞维安城墙（Servian Wall）为边界的罗马城区以外。这一状况直到公元 3 世纪下半叶的奥莱留城墙（Aurelian Wall）的建成才被改变。由于长期洪泛，这片平原一直作为罗马军队练兵场地之用，从而以战神马尔斯（Mars）命名。与此同时，随着古罗马广场山谷之地的建设日益拥挤，一些占地较大的公共建筑如跑马竞技场、剧院以及后来闻名于世

古罗马城区示意图

---

1 Sergei Eisenstein, "Piranesi, or the Fluidity of Forms", *The Sphere and the Labyrinth*, p.90.

2 John Wilton-Ely, *Piranesi as Architect and Designer* (New Haven: Yale University Press, 1993), p.40.

3 Jonathan Scott, *Piranesi* (London: Academy Editions and St. Martins Press, 1975), p.166.

4 Stan Allen, "Piranesi's Campo Marzio：An Experimental Design", *Assemblage* 10, p.70-109(72).

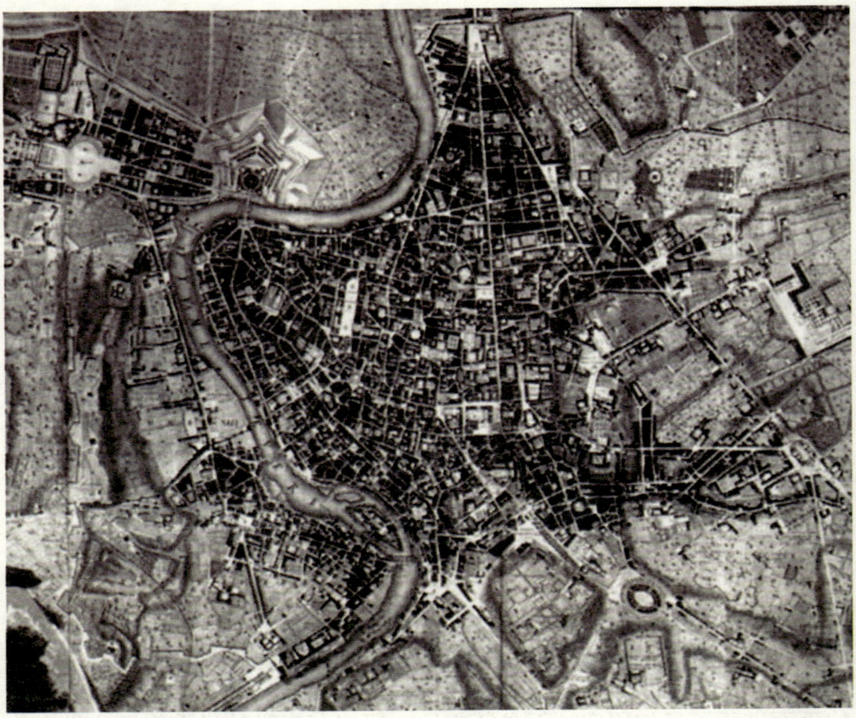

G.B. 诺利《新绘罗马地图》（1748）

的万神庙陆续出现于此。

另一方面，与他曾经参与其中的其他作品相比——比如 1748 年在 G.B. 诺利（Giovanni Battista Nolli）主持下完成的著名的《新绘罗马地图》（*Nouva pianta di Roma*），皮拉内西对“战神原”的“考古学重构”显示出非常不同的特质。在此，历史上业已发展成熟的罗马建筑类型的特定语言看似准确无误，但是不同建筑类型之间的组合却杂乱无章。纪念性建筑大量出现，其布局又显得漫不经心。建筑类型的变化可谓应有尽有，甚至到了肆意堆砌的地步。与诺利的《新绘罗马地图》相比，罗马城固有的街道和城市结构不复存在，既没有空间上的连续性，也没有结构上的整体性。即使在看似具有等级性几何结构的地区——比如在鸟瞰平面图上似乎占据着该地区中心的 S 弯河道地区，各自为政的建筑类型也使城市应有的结构原则荡然无存。皮拉内西的策略似乎是，首先将城市的形态语言推向极致，使之物极必反，然后再弃之不用。在皮拉内西的“考

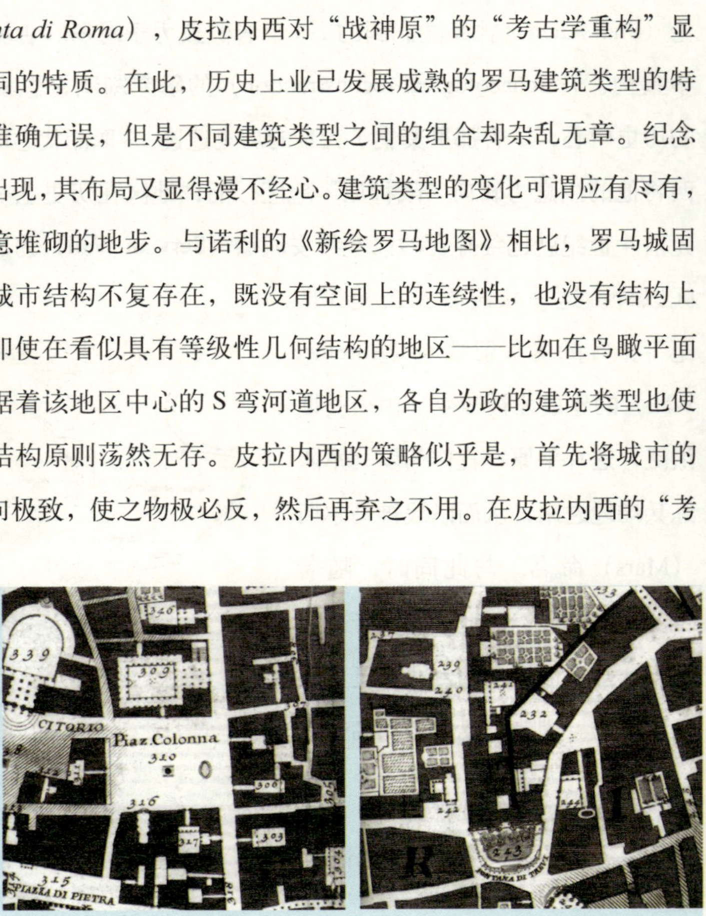

诺利《新绘罗马地图》局部

古学重构”之下，“罗马战神原”完全不是诺利的罗马地图所描绘的那种有特定边界且充满“聚集感”（gathering）的场所和城市；相反，它没有边界，仿佛正在向四面八方无限延展。

因此，无论是场地的选择，还是“重构”的结果，皮拉内西的《古罗马战神原》组图都显得非同寻常且别有用心。塔夫里的洞察一针见血：“皮拉内西《战神原》的考古学面具骗不了任何人：这是一个实验性的设计（an experimental design），城市本身仍然是未知的。”[1] 可以说，正是《古罗马战神原》组图奇异的“未知性”使它成为与罗马这个人们熟知的城市相对立的“非场所”（non-place）；也正是它的“未知性”使得皮拉内西的考古学重构项目成了场所概念的另一种“系统性批判”。当然，与《监狱》组画不同，《古罗马战神原》组图的场所概念批判很大程度上是在城市层面展开的，因而也是对城市概念的批判。

鉴于20世纪后期的城市状况，皮拉内西的《古罗马战神原》组图已经不再是一个不祥的“消极乌托邦”预示，而是当今世界许多城市情形的现实写照。哈佛大学教授亚历克斯·克里格（Alex Krieger）曾经如此将皮拉内西与当代北美城市相提并论：

> 如果我们观察大多伦多地区或者亚特兰大，或者波士顿历史城区以外的郊区，或者那些臭名昭著的边缘城市，比如弗吉尼亚州北部的泰森角（Tyson's Corner），我们看到的环境——无边无际和多中心、荒芜、过剩、偏离正轨、似乎从来没有“总体规划”、简直可以用一片混乱来形容——这难道不是与皮拉内西的罗马如出一辙吗？[2]

有趣的是，作为《小、中、大、特大》一千三百余页中为数不多的几个20世纪之前的建筑图像之一，库哈斯对皮拉内西的《古罗马战神原》图版的使用确实与当代城市有关，更确切地说与一篇题为《城市究竟发生了什么？》（“What Ever Happened to Urbanism”）的短文有关，而这个作为标题的问题正是提给当代城市的。就此而言，库哈斯与皮拉内西在历史和思想两个层面的亲缘关系似乎已经不言而喻。

二　库哈斯与“广普城市”的“偏执妄想的批判之旅”

在一部关于超现实主义的文集中，艺术史学家哈尔·福斯特（Hal

1 Manfredo Tafuri, *Architecture and Utopia: Design and Capitalist Development*, trans. Barbara Luigia La Penta (Cambridge, Massachusetts and London: The MIT Press, 1976), p.15.

2 Alex Krieger, "Conflicting Ideals for the City", *GSD News*, Winter/Spring, 1995, p.16.

哈尔·福斯特《强迫之美》英文版

Foster）描述了超现实主义者安德烈·布雷东（André Breton）于一次世界大战中在圣迪济耶（Saint-Dizier）的一家精神病诊所当助手的经历。据说，病人中有一位士兵坚持认为，这场战争子虚乌有，伤员是化妆的，死者是从医学院搬来的。这激起了年轻的布雷东的兴趣。在他看来，这是一位由于过度惊吓而进入“偏执妄想型另类现实”（paranoid alternative reality）的病人，而这种“另类现实”本身就可被视为对现实的某种批判。[1]我们无法准确知道圣迪济耶的这段经历对此后成为超现实主义者的布雷东究竟产生了多大影响，但是有一点可以肯定，偏执妄想现象对超现实主义的重要性无论如何强调都不为过。

确实，无论布雷东自己，还是乔齐奥·德·基里科（Giorgio de Chirico）、马克斯·恩斯特（Max Ernst），诸多超现实主义者都曾试图捕捉和探索偏执妄想对超现实主义艺术的意义。不过，诚如布雷东所言，在这方面一个更为关键的人物是萨尔瓦多·达利（Salvador Dali），正是他将所谓“偏执妄想的批判方法”（Paranoid-Critical Method）作为“最重要的工具”引入超现实主义。[2]在达利那里，“偏执妄想的批判方法”是一种“基于对谵妄的联想和诠释进行批判和系统的客观化而获取知识的自发方法”(the spontaneous method of attaining knowledge based on the critical and systematic objectification of delirious associations and interpretations）。[3]看起来矛盾的是，它在寻求“科学证据”的同时，提

1 Hal Foster, *Compulsive Beauty* (Cambridge, Massachusetts and London: The MIT Press, 1993), p.xi.

2 Andre Breton, *What is Surrealism?: Selected Writings*, ed. Franklin Rosemont (New York: Monad, 1978), p.136.

3 Dawn Ades, *Dali* (London: Thames and Hudson, 1982), pp.119-126.

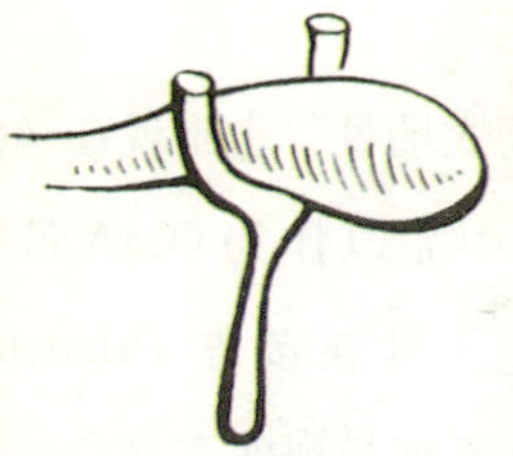

Dali's diagram of the Paranoid-Critical Method at work doubles as diagram of reinforced-concrete construction: a mouse-gray liquid with the substance of vomit, held up by steel reinforcements calculated according to the strictest Newtonian physics; infinitely malleable at first, then suddenly hard as rock.

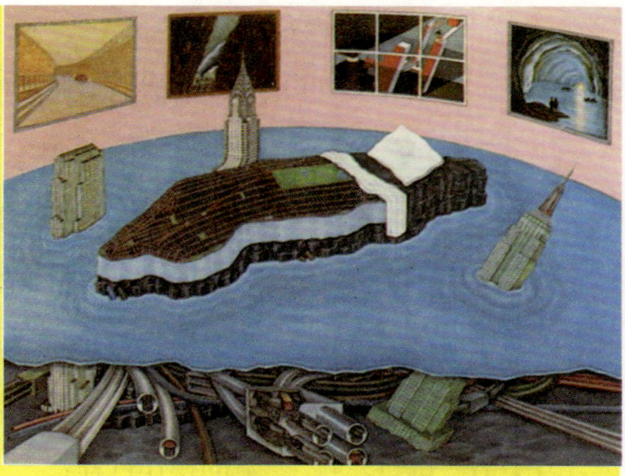

左：达利“偏执妄想的批判方法”图解　　右：维森多普 /OMA《无尽的弗洛伊德》

出了旨在“颠覆和质疑‘逻辑统治’和‘科学’确定性”[1]的“证据 + 妄想”的承诺，正是这一承诺带来了在“癫狂世界”中发现新的客观的“非理性意义”、寻求在“现实层面”进行创造性突围的可能。换言之，面对非理性和荒唐，“偏执妄想的批判”活动以癫狂的诠释为手段，将作为病理现象的偏执妄想转换成一种诗性发现的历程，最终在对现实素材的创造性想象中征服非理性。有趣的是，“征服非理性”（The Conquest of the Irrational）也是达利自己在论述“偏执妄想的批判方式”的文章中使用的标题。一定程度上，所谓“征服”可以用库哈斯引述达利“偏执妄想的批判方法”的图解时提供的说明进行理解：

> 达利的偏执批判方法图解实际上也可以理解为一个钢筋混凝土图解：一种由呕吐物组成的鼠灰色液体，搁置在根据牛顿物理学严格计算出来的钢架上面，开始还无法成形，却在突然之间坚如磐石。[2]

超现实主义和“偏执妄想的批判方法”是库哈斯建筑和城市思想的重要基础。从“大逃亡”开始，某种意义上的“偏执妄想的批判方法”就在库哈斯的作品中若隐若现。在此，弗里茨·纽迈耶曾经指出，一种大都市分裂现实的“弗洛伊德式场景”（Freudian tableau）跃然纸上，超现实主义的感知和再现“既使现实的饥渴找到了丰富而又奇异的养分，又以对矛盾的渴望在唐突的摩擦之中揭示出现实所包含的鲜为人知的诗

1 Dawn Ades, “Photography and the Surrealist Text”, in Rosalind Krauss and Jane Livingston, *L'Amour Fou: Photography and Surrealism* (London: Arts Council of Great Britain, 1986), p.159.

2 Rem Koolhaas, *Delirious New York: A Retroactive Manifesto for Manhattan* (Noew York: The Monacelli Press, 1994), p.248.

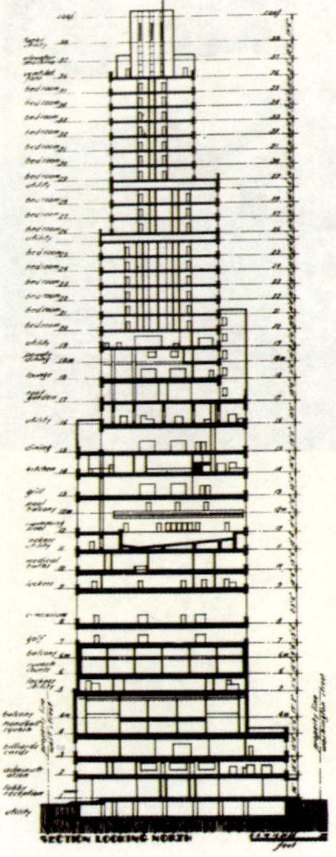

下城体育俱乐部剖面图

性内容”。[1]

类似的“弗洛伊德式场景”也出现在《癫狂的纽约》之中，其中尤以作为OMA最初四成员之一的玛德莲·维森多普（Madelon Vriesendorp）制作的一系列超现实主义绘画最为突出。与此同时，库哈斯将达利“偏执妄想的批判方法”概括为两个相互连续但又小心翼翼的操作过程：

> 1. 以新的眼光，综合再现偏执狂观看世界的方式——包括它极其丰富而又坚信不疑的对应关系、类比和模式；
> 2. 让这些似是而非的幻想的浓度压缩到一个临界点，使它们和事实一样密集：这一方法的关键是，捏造偏执之旅的实体化“纪念品”，或者说具体证据，再将这些出游的“发现”以旅游快照一般显著而又不可否定的理想形式反馈给全人类。[2]

从柏林到纽约，从亚特兰大到新加坡，库哈斯正是以“偏执妄想的批判方法”为手术刀，对20世纪世界都市之旅中出现的“癫狂”证据或者说“纪念品”进行剖析，得出既面对现实而又超越物质现象的认识。因此，尽管《癫狂的纽约》将“下城体育俱乐部”（Downtown Athletic Club）视为曼哈顿高层高密度城市模式的代表，所谓的“拥挤文化”（Culture of Congestion）却不能仅仅理解为物质层面的拥挤；相反，它首先是更为

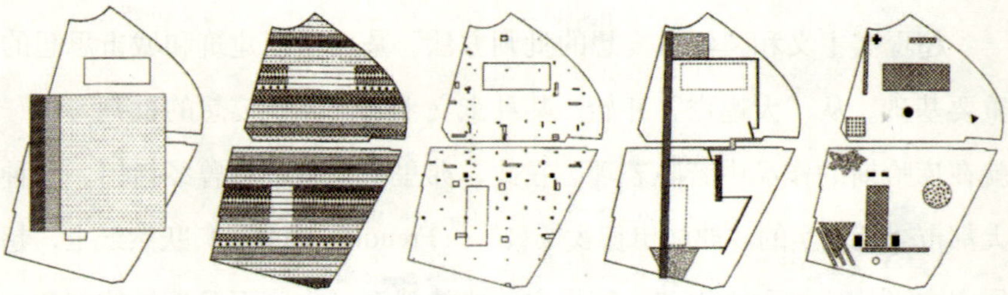

OMA／库哈斯：内容纲要的拥挤——巴黎拉维莱特公园设计竞赛方案图解（1982）

1 Fritz Neumeyer, “OMA’s Berlin: The Polemic Island in the City,” p.46 and p.38.

2 Rem Koolhaas, *Delirious New York*, p.238.

库哈斯：《给地球俘获的城市》(1972)

抽象的建筑和城市的“内容纲要的拥挤”（programmatic congestion）。正是这种“内容纲要的拥挤”成为库哈斯1983年的拉维莱特公园（Parc de La Villette）设计竞赛方案的核心。在《癫狂的纽约》中，“内容纲要的拥挤”也曾以“给地球俘获的城市”（City of Captive Globe）的面貌出现。在这里，大都市，或者说“拥塞文化”，采取了严格的曼哈顿网格形式，其中每个街块都俨然是一个自成一体的独立岛屿，最大限度地聚集着迥然不同的价值体系，无论这些价值体系是建筑学层面的，还是意识形态层面的，超现实主义、至上主义、构成主义、资本主义、勒·柯布西耶的光明城市，还有右下角那个柏林墙的缩影……

事实上，整部《癫狂的纽约》都是“偏执妄想的批判方法”无与伦比的建筑学产物，它将研究、佐证、历史、想象、虚构、谬误、巧合、超现实主义绘画、设计融为一体，并在1978年横空出世后大获成功。有趣的是，它在《小、中、大、特大》中仅仅与“大逃亡”一起占据了一个“前戏”（Foreplay）的位置。之后，“偏执妄想的批判方法”仍然发挥着这样或那样的作用，但是如果说在《癫狂的纽约》中库哈斯将自己设想为一个20世纪城市的“捉刀人”（ghostwriter）的话，那么正如《小、中、大、特大》显示的，他之后的写作则通过这种“捉刀”表现出对既有建筑思想的质疑和挑战。

大都会建筑事务所于1970年代中叶在纽约成立。在宣告成立的招贴画中，达利的影响再次昭然若揭：一个代表大都会建筑的塔楼从一个鸡蛋中破壳而出，令人想起达利1937年的画作《纳西索斯的变形》（*Metamorphosis of Narcissus*）中生长在鸡蛋上的水仙花。此后不久，库哈斯到伦敦的AA任教。库哈斯如此回忆当时建筑学面临的国际窘境：

左：OMA 成立宣言（1975）　　右：达利《纳西索斯的变形》（1937）

> 美国：后现代主义甚嚣尘上。欧洲：历史主义死灰复燃——这或许意味着“新”将永远被取代？美国：脱离文脉的自由。欧洲：文脉就是一切。美国：一切都大。欧洲：一切皆小。[1]

与此同时，大都会建筑事务所也搬到了荷兰的鹿特丹。与库哈斯的柏林一样，鹿特丹也曾经是一个历史中心，并且以两次世界大战期间出现的荷兰现代主义闻名于世。但是，这一切都在二战中毁于一旦。尽管城市在战后得到重建，却再也没有恢复到战前的城市状况，取而代之的是开放的城市中心，以及由荷兰现代主义建筑师巴克马（Jacob Berend Bakema）设计的著名的莱班（Lijnbán）商业街，并因此一度成为战后的模范城市。不过，这个开放的城市中心后来却成为鹿特丹的“大问题”。就像 1980 年代的柏林国际建筑博览会（IBA）倡导的那样，人们以城市重建的名义呼吁以封闭的街区填满开放的中心，美其名曰让城市更“城市”。可以说，鹿特丹的争议是“欧洲城市重建”（Reconstruction of the European City）思潮的一种反映。

库哈斯的立场与之大相径庭。对于库哈斯来说，不但这个城市独特的品质取决于其中心的开放性，而且如同柏林一样，原型序列的变化和破坏反倒蕴含着这个城市的丰富性。因此，如果说“欧洲城市重建”是

左：二战后的鹿特丹　　右：OMA／库哈斯　布姆派斯板楼设计方案（1980—1982）

1　S, M, L, XL, p.518.

以历史的名义抹去作为城市历史最重要的事实和记忆（第二次世界大战及其后续的冷战格局）的话，那么库哈斯看到的则是“20世纪令人惊骇之美”（the terrifying beauty of the twentieth century）。库哈斯指出，理解并欣赏这种“美”并非易如反掌，因为一旦秩序和完整性不复存在，“人们需要把夏普里昂（Champollion）[1]、施利曼（Schliemann）[2]、达尔文（Darwin）和弗洛伊德（Freud）融为一体，才能消化和理解随之而来的城市景观。”[3]

因此，这个所谓“20世纪令人惊骇之美”绝非仅仅是一个趣味或者审美的问题。相反，在看似趣味或者审美性的立场背后是库哈斯对方法的思考，它意味着对“偏执妄想的批判方法”的转化和延展。一定意义上，我们也可以将它视为《小、中、大、特大》的方法。库哈斯自己这样写道：

> 如果这部著作中存在一种方法，那么它就是系统的理想化方法（a method of systematic idealization）——就是将现实存在系统地推向极致，就是在思辨的狂轰滥炸中关注最平庸的现实层面，对其进行回溯性的概念和意识形态思考。每一个杂种都有一棵谱系之树；每一个最模糊的思想线索都被穷追不舍，就像充满奸情的侦探故事一样生动有趣。[4]

借助于这样的方法，OMA为鹿特丹设计的布姆派斯板楼（Boompjes TowerSlab）项目以及大致同一时期设计的柏林科赫大街（Kochstrasse）/弗里德里希大街（Friedrichstrasse）项目无疑构成了对1970年代后期建筑学面临的三大国际窘境中的第一个（后现代“历史主义”）的超越。与此同时，第二个窘境（文脉和非文脉）似乎没有得到解决。毫无疑问，这个意义上的“文脉”已经不再是柯林·罗式的“图形—背景”意义上

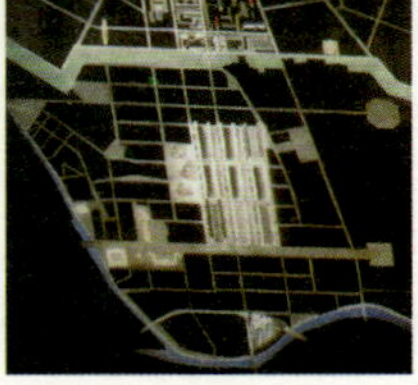

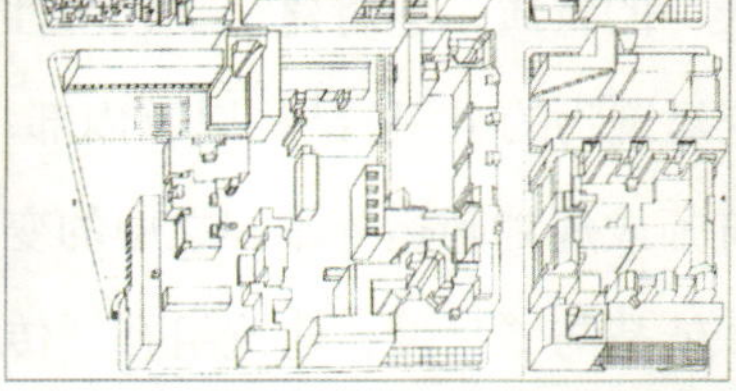

OMA/库哈斯(1980)　左：弗里德里希城区概念性鸟瞰　中：科赫大街/弗里德里希大街方案　右：公共住宅方案

1 让-弗朗索瓦·夏普里昂（Jean-François Champollion, 1790—1832），法国学者，埃及学创始人。

2 海因里希·施利曼（Heinrich Schliemann, 1822—1890），德国考古学家。

3 S, M, L, XL，p.206.

4 Ibid, p.208.

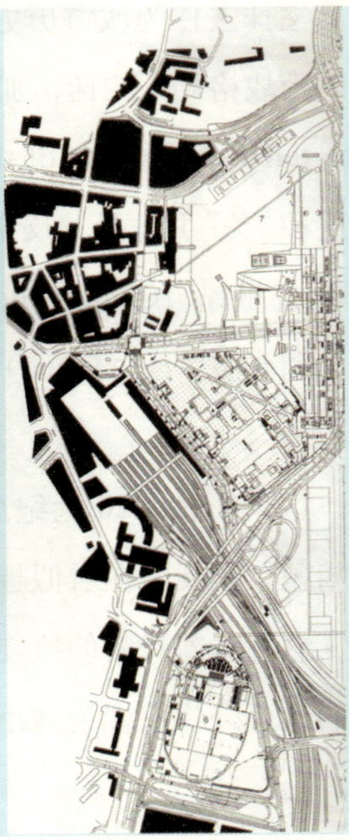

左：达拉瓦别墅（1985—1991）
右：欧洲里尔总平面（1988—1991）

的文脉主义（figure-ground contextualism）。但是“文脉”仍然可以被认为是一个基本的建筑学问题，或许因为任何建筑都有一个特定场地这样一个简单事实。场地条件可以根据形式或者空间肌理进行理解，也可以根据服务和供给配置、甚至力流（人流 / 车流）进行理解。就此而言，无论哪种情况，无论一座建筑物的环境是传统还是现代，场地都会对建筑产生影响。反过来说，对文脉的肯定也就意味着一种现实思考，即便这是一个混乱不堪且不可预测的现实，而直面这样的现实并由此激发想象在许多情况下似乎正是库哈斯策略的组成部分。

不同的是，如果任何对文脉的义务最终意味着一种改良主义立场，而这一立场使一切全新秩序（new orders *ex nihilo*）都无从谈起的话，[1] 那么库哈斯的态度从来都不是改良主义的。即使在混乱的城市中心，他也渴望虚无——“一切皆无之处，一切才有可能。建筑所在之处，一切皆无可能。”[2] 正是这种对虚无的渴望促使库哈斯试图以“大”（Bigness）的宣言超越文脉与非文脉的窘境。对库哈斯来说，一旦超越一定尺度，建筑就获得了一种“大”的特质，它将建筑“艺术性”和“道德性”统统扫地出门。“大”改变建筑物的组织、结构以及内部 / 外部的关系。“大”使传统意义上处在建筑师或规划师运筹帷幄之中的变得无法掌控。“大”抛弃城市文脉。“大”就是与“尺度、建筑组合、传统、透明性以及伦理的决裂”——最终，也是最根本的决裂：“大”不再是城市肌理的一部分。“大”只是存在；至多，它是一种共存。“大”的潜台词就是“去

1 Jacques Lucan, “Contextualism and Universality”, *Lotus International* 74 (1992).

2 S, M, L, XL，p.199.

XX的”文脉。[1]

“大”是反场所和反城市的建筑学宣言。在此，库哈斯不仅试图完全接受“大”对场所和城市的影响，而且将它作为一个有待探究的理论问题——也许，这就是库哈斯为“大”使用大写字母开头的 Bigness 的原因。通过“大”的宣言，库哈斯关注的不仅仅是建筑工程的大小或者尺度，其变化范围可以从小型的达拉瓦别墅（Villa dall’Ava）到欧洲里尔（Euralille）的巨无霸项目不等；他的兴趣也不仅仅在于超越文脉与非文脉的困境本身。要彻底打破文脉的命题，“大”的最终含义只能在一种新的城市概念的层面进行理解。

为寻求可能的答案，库哈斯再次赴美，他在《亚特兰大》（“Atlanta”）一文伊始开宗明义地写道：

> 有时，重要的是确定城市是什么——而不是它过去是什么，或者它应当是什么。这驱使我去亚特兰大——一种直觉，在那儿可以发现20世纪末真正意义上的城市……[2]

在库哈斯看来，亚特兰大并不是一个城市，而是一个广阔的区域。如同几乎所有美国城市一样，它曾经拥有一个“当城”（downtown）。但是，在1960—1970年代，当许多美国大城市的“当城”都陷于深深的绝望状态之时，亚特兰大的“当城”却被复苏，或者说彻底重建。当然，这并非通常意义上的城市重建，而是由建筑师和开发商双重身份的约翰·波特曼（John Portman）的更新计划促成的“一次处女重生：一次城市克隆”。整个过程从单个街区的改造开始，再通过人行天桥将街区连接起来，形成综合的城市系统。而这个体系的关键要素则是波特曼发明的建筑装置——“中庭”（Atrium），一个由巨大的玻璃屋顶覆盖的室内空间，四周环绕着酒店客房、写字楼、餐厅以及应有尽有的设施。每个中庭都像“当城”一样包罗万象——事实上，它就是“当城”的替代品。而且，随着中庭建筑相互聚集，“当城”仿佛被瓦解为许多“当城”，或者说城市就是由许多自成一体的“当城”形成的群体。

普拉德 & 库哈斯《亚特兰大》英文版（1999）

如此一来，看似旨在恢复和稳定亚

1 *S, M, L, XL*, pp.495-502.

2 Ibid, p.835.

特兰大"当城"的举措实际上却以两种方式加速了它的灭亡。首先，传统上作为户外空间的街道不再是一个有意义的概念：即使街道空间仍然存在，实际的城市生活都发生在建筑物内部。第二，"当城"被原子化，每个自成一体的粒子可以出现在任何地方：

> 现在到处都是城市。这个新的城市状况简直铺天盖地，把景观、公园、工业、锈蚀地带、停车场、住宅区、独户住宅、沙漠、航空港、海滩、河流、滑雪坡、甚至"当城"统统囊括其中。……亚特兰大不是一个城市，而是一个景观。[1]

亚特兰大的新城市景观——或者用库哈斯的术语来说，"后城市"（post-urban）景观——是"大"的卓越案例。库哈斯丝毫不想隐讳这个案例在 OMA 思想史上的重要性。他坦言，正是这个来自新世界的令人震惊的"大"，使得《癫狂的纽约》已经包含的某些内容变得明确。[2] 同样，美国建筑学者桑福德·昆特（Sanford Kwinter）也指出，没有什么像从纽约到亚特兰大的游历那样能够更好地说明库哈斯思想的发展：

> 在库哈斯那里，"美国"的概念一直在发展壮大。它不仅具有巨大的审美价值，而且也在 OMA 的作品（尤其是那些置身于欧洲城市语境的作品）中发挥了引发创新和激进色彩的作用。……对于库哈斯来说，对美国的深入研究已经成为他工作的一部分，并一直在策略上与欧洲保持着某种"危险的"——因此也是富有创造性的——距离：它已被设定为并巧妙地被保持为一扇粗陋的但又不无神秘色彩的大门，由此通向一个不稳定的但又可以激发创新的外部世界。[2]

如果库哈斯的世界城市之旅仅限于欧美范围，那么昆特的观察无疑完全正确。但是，在美国这个"新世界"之外，现代亚洲的发展提供了另一个新的地域，也为库哈斯探索"大"的理论提供了新的案例，并铸就了《小、中、大、特大》中的《新加坡歌集——一个波特金都市的肖像，或者是三十年的白纸策略》（"Singapore Songlines: Portrait of a Potemkin Metropolis ...or Thirty Years of Tabula Rasa"）一文。那么，新加坡在什么意义上成为"大"理论的试验场？当然不是在物理尺度或人口意义上的。毋宁说，"'大'，独立于文脉，它使建筑学得以在挖掘新的全球化'白

---

1 S, M, L, XL, p.850.

2 Sanford Kwinter, "Flying the Bullet, or When Did the Future Begin?", *ARCH+* 132, June-1996, p.80. 中文译文见：桑福德·昆特，《飞射子弹或者未来始于何时？》，载《莱姆·库哈斯与学生的对话》（莱斯大学建筑学院编，裴钊译），北京 . 中国建筑工业出版社，2003.

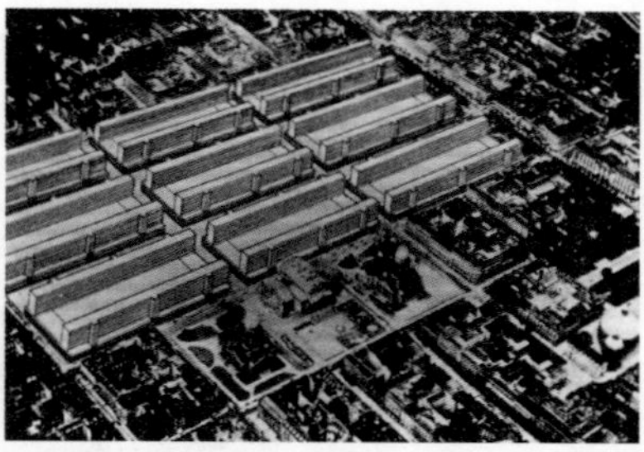

左：勒·柯布西耶《伏瓦生计划》（1925）　　右：希伯塞默尔的柏林市中心改造方案（1927）

纸策略'(tabula rasa)的条件下获得继续生存的机会。"[1]从历史角度说，"白纸策略"——或者说将城市的局部甚至整体作为一片净土重新开始——并非什么新的现代化现象。它首先发生在欧洲，从奥斯曼（Hausmann）的林荫大道、勒·柯布西耶的伏瓦桑规划（Plan Voisin）到路德维希·希伯塞默尔（Ludwig Hiberseimer）的柏林中心区的住宅项目，都是早期现代主义白板策略的不同案例。这种策略很快被视为现代主义的大恶。在这个问题上，库哈斯再次表达了他与既有建筑思想针锋相对的立场和观点，这导致他在 OMA 的巴黎拉德芳斯（La Défence）再开发设计竞赛项目中重访"白纸策略"，尽管与现代主义的形式偏好不同，它刻意采用了一个无形（formless）的变化形态，而且，尽管在欧洲语境中，该项目完全是理论性的，或者说是基于一个完全不现实的假设前提：

> 如果，即使在欧洲，或者说特别是在欧洲，我们宣布整个区域中凡是超过 25 年的建筑都不再有价值，变成一片空白，或者至少具有拆除的可能，那么情况又会怎样呢？[2]

但是，在东南亚，城市直接从 19 世纪飞进 21 世纪，这导致"白板策略"实际上被运用到史无前例的程度。那么，库哈斯发问，亚洲正在发生的一切是首先在欧洲打开的潘多拉盒子的始料未及的产物，还是亚洲的现

OMA／库哈斯：巴黎拉德芳斯改造设计咨询方案（1991）

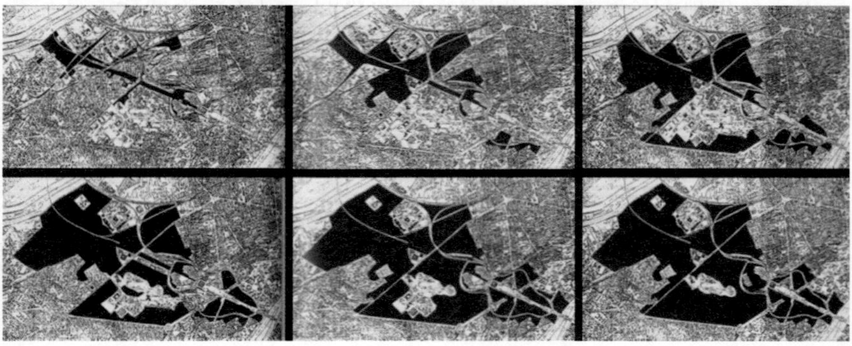

1 *S, M, L, XL*, p.515.

2 Ibid, p.1105.

新加坡　左上：公租房　　左下：亚洲乐园　　右：市中心

代化已经足够成熟，以至于能够超越城市传统与现代化的“两难”？正是在这样的意义上，库哈斯将新加坡视为“当代生态学”（ecology of the contemporary）的独特案例。[1]

诚然，30 年的“白纸策略”几乎使新加坡建筑文化的过去丧失殆尽，原来的村落现在只是一个全新世界中的残骸。然而，库哈斯指出，如果没有当地文化传统的支持，作为新加坡现代化进程之基础的彻底“白板策略”是没有可能得到贯彻的，这些文化遗产包括儒家传统以造福大众为名义对效率的厚颜无耻的追求，以及“封建残余与民主奶油”的混合。与此同时，“现代文化对更多自由的严肃要求”则被忽略甚至压制。[2] 就此而言，人们确实可以把新加坡视为基于文化遗产的现代化案例，尽管物质环境的面貌看起来与之截然相反。

另一方面，即使在“白纸策略”被推广到极致的 30 年中，新加坡寻求身份和特征的努力也似乎从未在物理环境的层面真正停止过。随之而来的是程式化的“亚洲”村落、中式园林、重建的各式寺庙、有“中式”屋顶的塔式建筑，以及遍地开花的后现代儒风商场。甚至一个村庄的异地重建也绝非匪夷所思。实际的情况也许是，“白板策略”的现代化进程并没有使文化遗产的物质形态和群族认同在新加坡灰飞烟灭；相反，它变得势不可挡，尽管这一切常常是以粗劣的历史赝品的面貌出现的。

从《新加坡歌集》到《广普城市》（“The Generic City”）只有一步之遥。换言之，“广普城市”的构建既是库哈斯的世界城市之旅在经历新加坡

1 *S, M, L, XL*, p.1011.

2 Ibid, p.1017 and p.1039.

之后的逻辑结果，也是“偏执批判方法”将以新加坡为雏形的东南亚城市之旅收集到的“事实”或者说“纪念品”压缩和转化为一种理想化形式的结果，是库哈斯将“出游的‘发现’以旅游快照一般显著而又不可否定的理想形式反馈给全人类”的又一次尝试。因此不奇怪，库哈斯的“广普城市”是多种族和多元文化的，它地处热带，商务活动清晰地体现在位于“当城”的摩天塔楼的形式之中。它的城市生活很大程度上集中在大型的购物中心。机场是城市大门，超级地方化和超级全球化的商品在这里应有尽有——事实上，机场就是一个无尽重复着相同结构单元的新型城镇。广普城市无形无状，它是道路、建筑物和自然的“自由式”组合。更重要的是，在这个“白板策略”生成的城市里，历史和身份之间相互依存的循环关系完全被打破：

> 一切只服从于眼前需要和能力。这是一个没有历史的城市。它足够大，每个人都能容纳于其中。它是轻松的，无需维护。如果太小，就自我扩张；如果太旧，就自我解构和更新。它处处令人兴奋，又单调乏味。它是“肤浅的”，就像好莱坞的摄影棚一样，每周一早晨都会提供一个新的身份。[1]

伊恩·奈恩（Ian Nairn）也许是对的，在经历了现代主义的“浩劫”之后，身份认同、特征、场所感都不是摆设，而是“脱离了它我们就无法生存的东西”。[2]“广普城市”在这一点上似乎也有过之而无不及。在那里，只要有一丝“身份认同”的线索出现，就会被无限放大，物尽其用到极致：

> 如果是滨水城市，那么以水为基础的符号就会遍布全城。如果是港口城市，船舶和码头吊架的形象就会一直延续到城市深处；如果是亚洲城市，那么“柔弱的”（感性和不可捉摸的）女性身姿就会无处不在，显示（宗教、性别）上的顺从，令人浮想联翩；如果城中有座山，那么每一个宣传册、菜单、门票都会重复山的主题，似乎不是天衣无缝的同义重复都不足以使人信服。[3]

这就是库哈斯通过“偏执批判方法”勾画的“广普城市”。它无情嘲讽了现代人对身份认同、特征和场所感不择手段的追求，因为在消费主义时代，一切最终又都将沦为赤裸裸的消费对象。它也表明，在一个消费文化当道的世界，对身份认同的焦虑反而是导致法国 20 世纪思想

1 *S, M, L, XL*, p.1250.

2 引自 Edward Relph, *Place and Placelessnes* (London: Pion Limited, 1976), p.32.

3 *S, M, L, XL*, p.1263.

家保罗·利科（Paul Ricoeur）所谓"最低劣的文化创造力"（the lowest degree of creative culture）的根源。有趣的是，就在警告"人类作为一个整体正日益趋向单一的世界文明，这一文明不仅意味着个体的巨大进步，而且也面临让我们的文化遗产继续生存并适应这一新条件的巨大挑战"[1]的同时，利科也指出一个"文化性创造的悲剧定律"（the tragic law of the creation of a culture）。之所以称之为"悲剧定律"，是因为真正的创造与身份认同的"虚假意识"（false consciousness）不可同日而语。说到底，它其实意味着对传统的反思和批判。在利科看来，创造性超越一切预设的定义，甚至"丑闻"（scandals）也成为必要。[2]或许，《小、中、大、特大》最终向我们展现的，正是作为这一"悲剧定律"有力印证的库哈斯自己的创造性理论话语和建筑实践。

结语

20世纪下半叶以来的场所和城市话语充斥着过去与现在的悖论，而人们在面对这一悖论时的价值偏向常常是在过去一边的。相比之下，库哈斯的理论话语和建筑实践展现了一条截然不同的思想路线。借助于塔夫里对先锋派与建筑的开创性研究，本文试图将这一思想路线追溯到18世纪的皮拉内西，从而为理解《小、中、大、特大》的理论话语寻求一个历史参照。皮拉内西《古罗马战神场》组图的考古学重构看似指向过去，但是在考古学面具的背后，他对创造性表达的渴望却导致了一个挑战建筑学和城市规则的实验性设计。简言之，它以挖掘"过去"为名，行"场所"和"城市"概念批判之实。一定意义上，《古罗马战神场》组图中的城市平面出现在《小、中、大、特大》中的《城市建筑学究竟发生了什么？》（"What Ever Happened to Urbanism"）一文的标题页并非偶然，而库哈斯的"广普城市"则可以被视为以"偏执狂批判方法"铸就的"20世纪考古学"（archaeologue of the 20th century），尽管这个"考古学"使用的工具既非皮拉内西的蚀刻画，更非传统考古学家的挖掘铲，而是成沓成沓的机票。很显然，库哈斯的"20世纪考古学"关注的是当代。通过观察和思考在一个快速变化的世界里场所或城市实际如何，他的理论写作是一套批判、超越、最终颠覆基于过去的"场所"和"城市"概念的话语。

但是，如果从皮拉内西到库哈斯的发展中存在一条思想路线的话，

---

1 Paul Ricoeur, "Universal Civilazation and National Cultures," *History and Truth* (Evanston: Northwestern University Press, 1965), p.271.

2 Ibid, p.281.

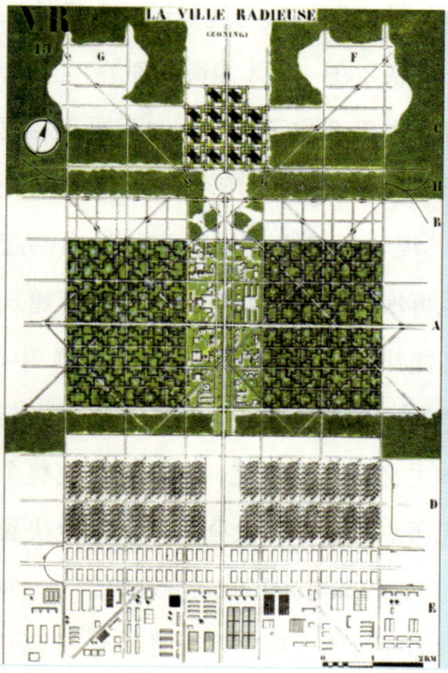

左：勒·柯布西耶《光明城市》（1935）
右：格奥尔格·格罗茨《弗里德里希大街》（1918）

那么他们之间的差异或许比相似之处更值得注意。皮拉内西用城市来做实验性设计，其主题用塔夫里的话来说就是“秩序的诉求与乱形的意志”（between the demand for order and the will to formlessness）之间的争斗。因此，现代主义一方面呈现为勒·柯布西耶“光明城市”严格的功能和形式秩序的诉求，另一方面又以“拼贴”（collage）和“蒙太奇”等艺术手段将“乱形美学”表现得淋漓尽致，从而在正反两个方面征服和化解现代主义在面对大都市前所未有的变革之时所产生的“焦虑”（anxiety）。对于库哈斯来说，20 世纪晚期的建筑和城市的新状况不仅改变了城市的概念，而且更重要的是它已经充分说明，城市不再是设计的产物，这使秩序和乱形的争斗失去意义。与皮拉内西的“古罗马战神场”不同，在库哈斯那里被关注的当代城市的杂乱无章和混沌无序并不是因为它们被设计成这样，而是诸多“不可视力量”（其中尤以资本的力量为甚）运作的结果。这使我们回到《小、中、大、特大》中那篇以《古罗马战神场》图版为首页的题为《城市究竟发生了什么？》的论文。事实上，也许没有什么比该文结尾部分典型的库哈斯式提问更能说明他与皮拉内西的区别：“如果我们直接认为，城市没有危机，并重新定义我们与城市的关系，不是作为它的制造者，而是作为它的主体，作为它的支撑者，那么情况又会怎样呢？”尽管该文的结语仍然宣称，“城市从未如此强烈地成为我们的全部所有。”[1]

因此，如果皮拉内西的“古罗马战神原”对场所和城市概念的颠覆应该置于启蒙理性批判的背景之下进行理解的话，那么在库哈斯那里，

1 *S, M, L, XL*, p.971.

塔夫里《建筑与乌托邦——设计与资本主义发展》英文版（1976）

问题的核心则必须与现代运动之后的认知转向联系在一起才能得到理解。但是，如同塔夫里开启了将皮拉内西置于启蒙理性的城市批判语境中进行研究的先例，[1] 现代运动之后的认识论转向的起点也再次可以追溯到塔夫里。在《建筑与乌托邦——设计与资本主义发展》中，正是通过对皮拉内西以来的先锋主义与资本主义发展的研究，塔夫里“使整个当代历史发生了一次强烈的认识论决裂（a sharp epistemological break），……他反对那些对建立和巩固所谓现代‘运动’的神话及其谱系津津乐道的历史话语，并以一系列颠覆性的问题将依附在建筑职业及学院期盼下的叙述化为泡影。”[2] 正是通过这样的认知转向，塔夫里才能够将现代建筑的意识形态和乌托邦欲望所面临的危机置于严厉的批判之中。按照塔夫里的观点，建筑师作为社会意识形态思想家地位的终结正是这场危机的结果，它最终说明，“‘设计的无所不能’只是一种陈旧的幻觉，一种既无能又无效的神话。”[3]

无论是否赞同塔夫里分析的政治立场，库哈斯的诉求实际上已经巩固为一种“后塔夫里”立场，同时也将塔夫里提倡的建筑学不应该接受既定现实条件的主张彻底放弃。事实上，库哈斯不仅接受而且热烈拥抱 20 世纪现代化和晚期资本主义全球化残酷无情的城市结果，同时完全避免了塔夫里在早期现代主义那里看到的焦虑和痛苦。他重温现代主义的“白纸策略”，但完全无意将其视为克服混乱和非可控力量的一种建筑手段，因而也抛弃了与现代主义“白板策略”如影随形的形式秩序和美学诉求。他相信一个前所未有的、无法预料的、从而也是具有巨大自由和潜能的城市状况才能提供建筑学专业浴火重生的可能。用库哈斯自己的话来说，“正由于失控，城市将成为想象力的主要源泉。重新定义之后，城市很大程度上不再仅仅是一种专业，而是一种思维方式，一种意识形态：接受现实存在。”[4] 很显然，这样一种无条件拥抱和接受现存的意识形态也意味着对一切意识形态的抛弃。诚如亚历杭德罗 · 塞拉（Alejandro Zaera）所言，库

1 Tafuri, “Reason’s Adventures: Naturalism and the City in the Century of the Enlightenment”, *Architecture and Utopia*, pp.1-40.

2 Jean-Louis Cohen, “Ceci n’est pas une histoire”, *Casabella* 619-620, 1995 January-February, pp.48-53 (49).

3 Tafuri, *Architecture and Utopia*, pp.181-182.

4 *S, M, L, XL*, pp.969-971.

哈斯倡导的是“一种战略性撤退，它放弃了对现代文明发展的意识形态抵抗”。[1]

鉴于当代文明和世界发展的诸多问题以及它的不确定性未来，也鉴于我们对现实的理解和处理必然充满价值判断，库哈斯的“不抵抗策略”本身必然会遭到反对和抵抗。然而，《小、中、大、特大》奇特的迷人之处似乎仍在于，它在将城市研究转化为“以完全理性的回答应对完全疯狂的问题”（perfectly rational answers to perfectly insane questions）[2] 的“快乐的科学”（a Gay Science）的同时，也使我们陷入这样一种无法摆脱的两难——借用尼克 · 兰德（Nick Land）在其著作《渴望毁灭——乔治 · 巴塔耶与恶毒的虚无主义》（*The Thirst for Annihilation: Georges Bataille and Virulent Nihilism*）中的对巴塔耶的评述来说，“接受他的写作乃是不可能，拒绝它们又只是将头埋进沙堆。人们处在骇异的兴奋之中，诚惶诚恐，却无处可逃”。[3]

后记 2016

本文最初以英文写成，并以《皮拉内西、库哈斯与场所概念的颠覆》（“Piranesi, Koolhaas and the Subversion of the Concept of Place”）为题发表于我曾经攻读博士学位的瑞典查尔姆斯理工大学（Chalmers University of Technology）建筑学院的建筑杂志《北欧建筑研究》（*Nordic Journal of Architectural Research*）1997 年第 2 期。“场所”是我博士论文《本质与文脉——一个关于场所概念的研究》（*Substance and Context: A Study of the Concept of Place*）的研究主题。一定意义上，这篇文章是我博士论文研究的延续和发展。

该文发表的时候，我已在同济大学建筑与城市规划学院。当时库哈斯在国内建筑界还是鲜为人知，更不要说对《小、中、大、特大》这本书的认知和讨论了。一眨眼 20 年过去了，一切皆今非昔比。首先，库哈斯已经成为中国建筑界家喻户晓的名人，他的《建筑素描》（*EL croquis*）专辑成为年轻建筑师和学子学习乃至追捧的对象。其次，不仅（*Mutations*）和《大跃进》（*Great Leap Forward*）都是库哈斯带领哈佛大学设计研究生院学生研究珠江三角洲当代城市化的成果。当然，

1 Alejandro Zaera, "Notes for a Topographic Survey," *El Croquis* 53, OMA/Koolhaas 1986-1991, March 1992, pp.32-51 (50).
2 Koolhaas, “Understanding the New Urban Condition: the Project on the City”, *GSA News*, Vol. Winter/Spring, 1996, p.13.
3 Nick Land, *The Thirst for Annihilation: Georges Bataille and Virulent Nihilism* (London: Routledge, 1992), p.xii.

OMA/ 库哈斯《满溢》英文版（2004）

他也完成了《购物指南》（*Guide to Shopping*）、《满溢》（*Content*）[1]、《拉各斯》（*Lagos*）等看似与中国没有直接关系的著作。由于出版时间的原因，本文没有可能涉及后面这些著作，然而本文讨论的库哈斯思想的某些特质仍然清晰地贯穿在这些著作之中：全球视野（世界城市之旅）、偏执批判方法、接受现实的主张、对主流建筑学思想的挑战以及改造，等等。

在所有这些特质中，最具争议性的、当然也是库哈斯思想中最具挑战性的观点无疑是接受（甚至是所谓“无条件”接受）现实的主张。但是，正如本文最后指出的，鉴于当代文明和世界发展的诸多问题以及它的不确定性未来，也鉴于我们对现实的理解和处理必然充满价值判断（如果不说意识形态的话），库哈斯的“不抵抗策略”必然会遭到反对和抵抗。事实上，就在库哈斯从欧洲到北美再到亚洲的世界城市之旅进一步发展至非洲，并在拉各斯极端混乱的城市状况的记录中找到表达之时，琼·奥克曼（Joan Ockman）曾经这样尖锐地写道：“这座地球上最地狱般的尼日利亚城市，面对这本关于它的病态混乱的审美化相册中所呈现的景象，人们不得不畏缩。”[2]

另一方面，大约在2000年前后，原本还是在库哈斯的思想和写作中若隐若现的对“批判性建筑学”（critical architecture）的质疑和批判被演化为一场以美国东海岸主要建筑学院为阵地的“批评”与“后批评”（post-critical）或者说“批判性建筑学”与“投射性建筑学”（projective architecture）的争论，以及《时代建筑》杂志对这场争论的代表性文献的系列翻译和介绍。库哈斯并非这场争论的主角，但是诚如乔治·贝尔德

---

1 书名使用的Content在中文中有“内容”和“满意”两种意思，这里基于该书第91页content的词条解释将其译为“满溢”：If a building or building-idea or book has content, conveys content, disseminates content, then it is a container: in other words, it is close to nothing. It is certainly no organism, nothing with signs of life. It is defined by its empty capacity, like the massive nondescript atrium of the Hotel Interchaneable, the space designed to impress rather than to live and give life. Artists lost something when the market made them content providers. OMA offers its Content and its content under full and cognizant eraure. We stress, again, that this list is an act of impeachment, not conviction: authority in language belongs to the open forum, not the self-appointed arbiters. And, having sounded an awakening bugle-blast, we're more than content to let reveille turn into jazz, step aside for someone else's solo, and let the cleasing jam session begin. 换言之，“满溢”试图表达上述词条解释在“内容”和“满意”两个含义层面上的使用以及由此而来的双关性：“满溢”在发音上与“满意”相同，同时在字面上又意味着某种“内容”的溢出，但是因为溢出，显然比固定的内容更多，而且溢出后也更不确定。一定意义上，“满溢”不仅符合该书刻意营造的内容满满以至于横流四溢的形象，而且也符合上述词条解释中对空洞的形式（container, content provider）和趣味准则的否定和批判。当然，“满溢”只是对该书书名的翻译，不应该取代上述词条解释中“内容”和“满意”的字面使用。

2 琼·奥克曼，《¥€$ 人》（王颖译，刘宇扬校），《时代建筑》2006年第5期（总91期），第43页。

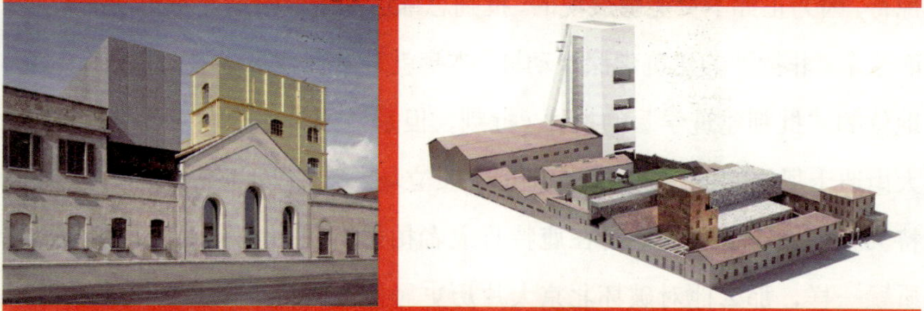

OMA／库哈斯：普拉达基金会保护性改造

(George Baird）所言，库哈斯在1994年加拿大建筑中心的ANY会议上提出的观点就已经显示了这场争论的“前兆”。“当前建筑批评中的一个问题”，库哈斯在发言中指出，“就是没有认识到，建筑最深层的动机中包含有非批评的东西”[1]，而这个“非批评”的“深层动机”大概就是奥克曼引述的库哈斯名言“只有面对现实，我们才能够采取行动”[2]所要表达的。

值得注意的是，尽管认识到库哈斯的“非批评的建筑学深层动机”说对这场争论的“前兆”作用，从而在一定程度上推动了“批判性建筑学”统治地位的溃败，但是贝尔德也敏锐地意识到库哈斯是“强劲的批评活动”的参与者，比如对大面积迪士尼化而导致曼哈顿42号大街独特街区文化的破坏的批评，或者对中国权威机构的批评，指责他们允许对北京大片历史居住区的破坏是可悲的、过分实用主义的立场，等等。[3]2004年，他在哥伦比亚大学的讲座中语出惊人：自米兰普拉达基金部总部(Fondazione Prada）改造项目以来，“保护正在成为我们的压倒性关切”（preservation is overtaking us）。2016年美国建筑师协会（AIA）年会之际，库哈斯在与莫森·莫斯塔法维（Mohsen Mostafavi）的最近访谈中这样表明自己对“保护”（preservation）问题的关注：

> 保护的美妙之处就在于你从已经存在的事物开始，因此也是从当地的事物开始。从定义上讲，保护性工程是对早期文化和状态的一种致敬，同时你为之增加新的维度、新的功能、新的美或者说感染力。各种信号都表明，需要对全球化重新思考并进行调整。[4]

1 乔治·贝尔德，《“批判性”及其不满》(都铭译，朱剑飞校），《时代建筑》2007年第3期（总96期），第55页。琼·奥
2 克曼，《¥€$人》，第43页。
3 乔治·贝尔德：《“批判性”及其不满》，第57页。
4 http://www.fastcodesign.com/3060135/innovation-by-design/rem-koolhaas-architecture-has-serious-problem-today

也许，人们不会对贝尔德所谓的库哈斯的“强劲的批判活动”感到惊奇，因为正如本文能够展现的，批判思维历来是库哈斯建筑思想的核心，而这个“非批评的建筑学深层动机”本身就是对在建筑学界长期占“主流”地位的“批判建筑学”的批评 / 批判。但是，如果库哈斯对曼哈顿 42 号大街迪士尼化的批评完全符合他的一贯立场，就如人们在《小、中、大、特大》中能够看到的他对在鹿特丹或者柏林出现的“欧洲城市重建”的质疑一样，那么他对破坏北京大片历史居住区的批评以及他近来在“保护”问题上的高调姿态则确实与“大”“去 XX 的文脉”“重温白板策略”的主张相去甚远。

应该如何理解这样的矛盾？很容易将库哈斯视为玩世不恭，阴柔巧妙，浑水摸鱼，欺世盗名的化身。事实上，在过去的 20 年间，对库哈斯的这类批评比比皆是，而且情况在国内更甚。特别是库哈斯设计的央视大楼因为《满溢》最后的几张图片而被冠以所谓“色情门”事件之后，对库哈斯的谴责之声更是不绝于耳，几乎使他成为全民公敌。[1] 不同于这种理解，本文仍然将库哈斯视为一个严肃的思想者——当然这样做绝不应该意味着对他的“迷信”[2]。事实上，“严肃的思想者”常常充满矛盾——柯布、福柯，无不如此。但是，如果柯布和福柯与库哈斯是可比的，那么我更愿意将库哈斯视为建筑界的福柯，而不是当今的柯布。这是因为，在柯布的思想中“永恒”仍然不可或缺，[3] 而与福柯一样，[4] 库哈斯的建筑思想则体现了一种对“确信”的绝对否定。一定意义上，正是这一特质导致了库哈斯思想的流动、多变甚至是对立。

也许，人们可以将库哈斯的思维方式概括为总是在既有（常常是主导的）思想和实践模式之外寻求新的可能，而这往往与库哈斯提出的观点所针对的语境紧密相关。“大”“去 XX 的文脉”“重温白板策略”针对的是这些论点提出之时在欧美建筑学盛行的后现代历史主义思想和实践。在这方面，或许没有什么比库哈斯对鹿特丹和柏林中心重建的后现代历史主义策略的质疑更有助于我们全面理解什么是真正的历史文脉——就此而言，自米兰普拉达基金会总部改造项目以来库哈斯在“保护”问题上的策略与他对历史文脉的态度并没有本质不同。另一方面，也没有什么比一个完全假设性的“重温白板策略”的拉德方斯改造方案更能体现库哈斯在欧洲城市更新的主导模式中寻求其他可能的渴望。同样，

1 见萧默：《CCTV 总部与臀部的“异质同构”！》（https://www.douban.com/group/topic/7059394/）。也见金秋野：《库哈斯嘲弄了谁》，载《读书》2009/11, pp.77-80. 需要看到的是，在《满溢》看似放荡不羁且充满反讽和黑色幽默的面貌之后，其实是库哈斯对建筑学及其当代条件的相当严肃的思考，可惜这些思考都因对“色情图片”的过度关注而忽视了。

2 参见金秋野：《不可迷信库哈斯》，载《读书》2010/3, pp.101-104.

3 参见本文集《勒 · 柯布西耶 *Vers une architecture* 译名考》。

4 关于福柯的这一认识，见杜小真：《选编前言》，载《福柯集》，上海：上海远东出版社，2003 年版，第 1-5 页。

亚特兰大、新加坡、拉各斯、"广普城市"的研究试图展现的无疑也是"另类"城市实际甚至虚构存在的可能。

然而，在经历了过去30多年中国城市以政府大规模卖地和拆迁为主导模式的改造历程之后，任何一个具有批判思考的建筑师/非建筑师大概都不会对这样的主导模式熟视无睹而不产生质疑，而库哈斯对破坏北京大片历史居住区的批评则可以作为这种质疑的一部分。它涉及的基本问题无疑是，中国的城市更新难道就只有这一条道路而没有其他可能吗？在这样的意义上，库哈斯的"新加坡"以及以此为雏形的"广普城市"既是一种印证，也是一种反讽。作为印证，它在太多方面呈现了过去30多年中国城市发展的道路和模式：国家主导下的"白纸策略"、大拆大建、城市生活由街道转向购物中心包罗万象的室内空间、物质性历史遗产破坏后的假古董和迪士尼化的猖獗盛行、以历史文化身份和物质福利压制现代化对真正自由的诉求，等等。作为反讽，它也说明，在一种语境（欧洲语境）中看似"离经叛道""奇特别致"的"另类"发展之路，在另一种语境中却是亟待反思和批判的对象。在后一个方面，除了奥克曼指出的道德悖论之外，库哈斯的城市研究和城市思维的不足之处也许还在于它们常常过于"大叙事"（如果不能说"宏大叙事"的话）。应该看到，作为城市研究，他的"偏执批判之旅"尽管不乏睿智的观点和灵光一现的洞见，但也常常有失于浮光掠影，甚至有猎奇之嫌，"亚特兰大""新加坡""拉各斯"如此，《突变》和《大跃进》也不能幸免。

然而，无论我们可以在什么层面保持对库哈斯城市话语的质疑，以"广谱城市"的"偏执批判之旅"为载体的库哈斯式的理论思维以及作为其集大成者的《小、中、大、特大》的力量仍然在于其巨大的批判性质疑所体现的反思能力。或许，我们可以采用迈克尔·斯皮克斯（Michael Speaks）的术语，将这种反思能力称为"智慧"（intelligence）。[1] 在我看来，正是这种"智慧"使库哈斯一直处在当代建筑思想既充满争议又不无突破的地带。同样，也是这种"智慧"使库哈斯的城市研究与他的建筑实践保持着一种特殊的关系。换言之，与其将库哈斯的城市之旅视为道德沦丧之地，或者以更为肯定的眼光将其视为建筑设计概念的生成器（比如从下城体育俱乐部的剖面到拉维莱特公园设计竞赛方案总平面的演变所显示的那样），不如作为铸就"智慧"能力的演练场。正是这种"智慧"能力（而不是某个具体的可以直接拿来用的设计概念）为库哈斯的设计提供了滋养的资源。也许，这就是库哈斯自己所言"在我所

1 迈克尔·斯皮克斯：《理想、意识形态、实用智慧——在中国与西方》（吴名，谢诗奇译），《时代建筑》2006年第5期（总91期），第63-64页；《设计智慧与新经济》（凌琳译，王骏阳校），《时代建筑》2007年第4期（总96期），第80-83页。

写和我所做之间，有一个巨大的、谨慎的、我认为是健康的出入”（There is an enormous, deliberate, and – I think – healthy discrepancy between what I write and what I do）的含义吧。与此同时，在“参数化”和“算法主义”（algorism）大行其道的今天，也许正是这样的“智慧”可以为建筑学抵抗这一趋势的一统天下保留某种希望。

10

Piranesi, Koolhaas and the Subversion of the Concept of Place[1]

1 First published in *Nordic Journal of Architectural Research*, 1997, No. 2, and slightly revised in this edition.

Among contemporary architectural thinkers, Rem Koolhaas is certainly one of the most intriguing. He persistently challenges the lines of thinking that architects and planners employ in the name of human scale, historic values, and the like. He dismisses the assumption that place and identity are necessarily dependent upon the past, whether this necessity is argued in terms of the Heideggerian sophistication or popular consensus. He embraces without reservation the forces that shape 20th century urban civilization in an attempt to explore the remaining possibilities by accepting and even applauding much of what others regard as the failure of 20th century urban development:*tabula rasa*, large-scale planning, and the consequent loss of character and identity of place. His strategy is to discern the sublime in the vulgar, hope in the terrifying, reason in the schizophrenic when it comes to urbanism of our time.

S, M, L, XL, an *magnum opus* over one thousand three hundred pages by OMA, Koolhaas, and Bruce Mau, first published in 1995 by the Monacelli Press, is a spectacular manifestation of this. A conglomeration of essays, fictions, diaries, travel logs, projects, drawings, models, photos, cartoons, and newspaper ads, the book shows not only the miscellaneous forms of Koolhaas' practice as an architect, but also the imprint of his earlier career as a journalist and screenwriter. Here, under a title borrowed from garment size, writing and project materials are organized from small to extra-large either according to the scale of building construction or in terms of the scope of the thinking — or rather the magnitude of its subject. However, as the authors suggest, the book has no connective tissue binding one part to another. Many of the writings are embedded between projects as autonomous episodes rather than supportive mortar. Taken together, the book is an outcome of the authors' striving for the presentation of a non-linear and even chaotic condition in the same way that "architecture is by definition a chaotic adventure."[1]

At this point one is reminded of the affirmative "nomad thought", exercised by such thinkers ranging from Friedrich Nietzsche to Gilles Deleuze and Félix Guattari. Nomad thought

1 O.M.A, Rem Koolhaas and Bruce Mau *S, M, L, XL* (New York: The Monacelli Press, 1995), p.xix.

operates in smooth, open-ended flowing spaces where one can rise up from any point and move to any other, free from codification whether ideological, institutional, or professional. Nomad thought derives from the vagabond imagination of the savage heart on a constant inner and outer voyage, aspiring to transcend the constraints of experience and thought.

Koolhaas' inner voyage parallels his outer one: the book can be read as a documentary of his tour of the world over the past decades, and of his inquiry into the condition of 20th century architecture and cities under the impact of politics, economics, and globalization. Although he first became known as the author of very much celebrated *Delirious New York*, the starting point of Koolhaas' journey and thus OMA's place of origin is, as Fritz Neumeyer has suggested, "not New York, as one might assume, but Berlin."[1] For as early as a student at the Architectural Association in London, driven by "intuition, unhappiness with the accumulated innocence of the late sixties, and simple journalistic interest"[2], Koolhaas chose "The Berlin Wall as Architecture" as the theme for his final thesis project. One year later, in 1972, in an entry for the *Casabella* competition on the theme of "A City with a Significant Environment" in which he collaborated with Elia Zenghelis, the thread of the *Berlin Wall as Architecture* reappeared, but this time took the form of fiction about the city of London. Entitled *"Exodus, or the Voluntary Prisoners of Architecture"*, the entry gives a direct allusion to the Berlin Wall: "Once, a city was divided in two parts. One part became the Good Half, the other part the Bad Half. The inhabitants of the Bad Half began to flock to the good part of the divided city, [their flight] rapidly swelling into an urban *Exodus*."[3] The story describes an "artificial paradise", a strip of land that runs through the center of London, "protected" from the existing city by two walls along its perimeter. Inside, the zone is subdivided into a series of identical squares, each with its own program, ranging from private allotments to communal facilities. Together, these squares are to restore the debased ideals of the metropolis "to a sparkling intensi-

1 Fritz Neumeyer: "OMA's Berlin: The Polemic Island in the City," *Assemblage* 11, pp.37-53(38).

2 *S, M, L, XL*, p.216.

3 "Exodus/1972," *Architectural Design* 47,1977, No. 5, p.328.

ty that would tempt the inhabitants of the subconscious London to escape into the strip in an impulsive *Exodus* — and to become its Voluntary Prisoners."[1]

Depicting a "paradise" which is "good" enough to attract the inhabitants of London and thus turns the physical structure of the old city into a pile of ruins, *Exodus* is certainly utopian in its nature. However, as Demetrios Porphyrios observed early in the 1970s, the utopian depiction of Koolhaas' and Zenghelis' *Exodus* does not aim to expose the ills of contemporary cities as its historical predecessors did, nor does it propose solutions to cure or redeem those ills.[2] Instead, the aim of *Exodus* is to awaken the sleepwalking metropolis of London and to insert in its inarticulate organism a social condenser of "totally desirable alternatives". At the same time, as its "architectural warfare" stems from the "hedonistic science of designing collective facilities that fully accommodate individual desires"[3], it is fraught with scenarios as from George Orwell's *1984*, the purgatory-like reception area, the totalitarian supervision of the allotments, and so on. Thus, the idealized metropolitan prototype, the "Good Half" of London, turns out to promise nothing but relentless pictures from a doomed civilization, an *apocalipsis cum figuris*. Or, rather than a utopia conceived on "goodness", *Exodus* is wrought with "dirty realism" in a horrifying psychological confrontation with the Berlin Wall emblematic to the delirium and miseries of the 20th century metropolis.

From the very beginning of his architectural career, therefore, Koolhaas has abandoned the intellectual foundations of a "polite" architecture. In its place, one finds a "negative utopia" that eliminates the distance between imagination and annihilation, between totality and nothingness, and between utopia and anti-utopia. Throughout architectural history, only a few have approached architecture in such a way. One of them was undoubtedly the eighteenth century Italian architect and etcher Giovan Battista Piranesi whose *Campo Marzio* plate is one of few pre-20th century architectural images that have appeared in *S, M, L, XL*.

---

1 Ibid.

2 Demetrios Porphyrios, "Pandoras Box," *Architectural Design* 47, p.357.

3 Ibid.

Piranesi and the "Archaeological Reconstruction" of the *Campo Marzio*

To my knowledge, Manfredo Tafuri was the first to expound on Piranesi's work in relation to the avant-garde and modern architecture of the 20th century. In *The Sphere and the Labyrinth*, Tafuri starts with a chapter entitled "'The Wicked Architect': G. B. Piranesi, Heterotopia, and the Voyage". To begin an analysis of the relationship between the avant-garde and the architecture of the 20th century with Piranesi is, as Tafuri acknowledges, undoubtedly provocative.[1] To a certain degree, what makes this historical analysis of the avant-garde possible are the frequent references to the works of Piranesi in the writings of Sergei Eisenstein, one of the leading film makers of the modernist avant-garde. Tafuri refers to Eisenstein's exceptional study of Piranesi's work *Le Carceri* as the clue to indicate the origin of the avant-garde with Piranesi. It is with his *Carceri*, Tafuri suggests, that the theme of imaginary sceneries with all its ideological significance enters into the history of modern architecture. Thus, Tafuri logically continues his study with the theme "the stage as 'virtual city'" in which the field of the avant-garde expands from the cabaret to the metropolis.

Tafuri's introduction to *Carceri* is particularly interesting for, as he points out, "what must be made clear from the start is that all this breaking up, distorting, multiplying, and disarranging, apart from the emotional reactions it can elicit, is nothing more than a systematic criticism of the concept of place, carried out by using the instruments of visual communication."[2] For, if the concept of place is inherently associated with the center, then *Carceri*, with its constant metamorphosis of space, is a "systematic criticism of the concept of 'center' "[3]. If the concern for the concept of place has resulted in the need for a clearly articulated language of architecture, then the ecstatic intertwining of forms and the undermining of the laws of perspective of *Carceri* present an "unequivocal attack on the 'language' insofar as it is a mode of acting upon the world."[4] And if the legitimacy of the concept of place is justified

1 Manfredo Tafuri, *The Sphere and the Labyrinth: Avant-Gardes and Architecture from Piranesi to the 1970s* (Cambridge, Massachtts and London: The MIT Press, 1987), p.55.

2 Ibid. 27.

3 Ibid.

4 Ibid. 34.

in terms of the subject's dependence on history and nature, then, in the "negative utopia" of *Carceri*, "History and Nature become detached from the subject, not to open up a new universe of values, but rather to present this radical divergence as the only possible value." [1]

Much of Tafuri's analysis of *Carceri* holds true for another of Piranesi's works, *Il Campo Marzio dell' Antica Roma*, published in 1761-62 shortly after reworking and elaborating of the *Invenzioni Capricciose di Carceri.* But whereas the "systematic criticism" of the concept of place in *Carceri* appears as a message from nowhere, *Campo Marzio* brings us straightforwardly to Rome, the most sovereign city. Unlike *Carceri*, in other words, *Campo Marzio* seems to hold a historic gesture, carrying on his earlier archaeological investigation of Roman architectural and engineering achievement in *Prima Parte di Architecture e Prospettive*, and *Opere Varie di Architettura, Prospettiva, Groteschi, Antichità*, and further in the four volumes of *Le Antichità Romane.* On the other hand, different from the frenzy, or to use Eisenstein's word, " the 'ecstatic' method of constructing"[2] of *Carceri*, the archaeological approach of *Campo Marzio* seems rational, showing a "scientific" investigation of the history of Rome, "supported by a sound scholarly method, appropriate to the Enlightenment and its concern with the phenomena of historical change." [3]

However, as Stan Allen has noted, in a project ostensibly devoted to the "archaeological reconstruction" of ancient Rome, Piranesi clearly ignored the historic, monumental center of Rome, where the existing ruins were concentrated and stood more or less free of contemporary building.[4] Instead he turned to an area called Campus Martius in Latin, or *Campo Marzio* in Italian, an area that remained outside the boundary of Rome — the Severan Wall — until the Aurelian Wall was built in the second half of the third century. Originally associated with the training of the young and with military exercises, and later open to pleasure and spectacle and even to funerals and burials, *Campo Marzio* was characterized not only

1 Ibid. 38.

2 Sergei Eisenstein, "Piranesi, or the Fluidity of Forms," appendix, *The Sphere and the Labyrinth*, p.90.

3 John Wilton-Ely, *Piranesi as Architect and Designer* (New Haven: Yale University Press, 1993), p.40.

4 Stan Allen, "Piranesi's Campo Marzio An Experimental Design," *Assemblage* 10, p.72.

by its marginality, but also by its otherness, by its incongruity with the city proper of ancient Rome.[1] What is more, in this "archaeological reconstruction", while the historically developed language of specifically Roman building types is affirmed, the combination of individual building types appears by and large disorderly. Monuments are massively presented but arbitrarily disposed with the same intensity as building types vary for their own sake. Streets are in the state of not existing, as has the entire domestic fabric of the city neither in terms of spatial continuity nor structural integration. Instead, what is at work is a hodgepodge of individual wills resulting in a heterogeneous void, a chaotic, or even senseless heap of building objects that have little to do with one another. Even in the areas composed with geometric and hierarchic structures, such as the double bend of the river where the plan and bird's eye view seem to represent the heart of the whole area, the extravagant diversity of building types undermines the principles of urban organization. A language of urban form is thus first exaggerated and then transgressed. Altogether, *Campo Marzio* gives a strong impression that violates the very notion of the city that previously was depicted by Giovanni Battista Nolli's *Nuova Pianta di Roma* from 1748.

Inasmuch as Piranesi grounded his reconstruction on archaeological and topographical studies of the site including the evidence provided by the fragments of the *Severan Marble Plan*,[2] also known as the *Forma Urbis Romae*, a fragmentary marble map of ancient Rome dating to the reign of Septimius Severus (circa 200 A.D.) and probably the single most important document on ancient Rome, *Campo Marzio* cannot be regarded as a product of its author's pure imagination. Piranesi himself stressed this point in a letter to the Scottish designer Robert Adam, saying that, in combining his on-site investigations with a careful study of the old plan of the city in the Capitol and the knowledge of the best historians of the time, his ultimate goal was that "no one would claim that I had followed my own whim rather than ... any evidence."[3] Meanwhile, Piranesi made no secret his aspiration to move beyond the established lan-

1 Ibid. 75.

2 John Wilton-Ely, *Piranesi as Architect and Designer*, p.40.

3 Jonathan Scott, *Piranesi* (London: Academy Editions and St. Martins Press, 1975), p.166.

guage of architecture as well as the city:

> But before anyone accuses me of falsehood, he should, I beg, examine the ancient plan of the city which I have just mentioned, he should examine the villas of Latium and that of Hadrian at Tivoli, the baths, the tombs and other ruins, especially those beyond the *Porta Capena*, and he will find that the ancients transgressed the strict rules of architecture just as much as the moderns. Perhaps it is inevitable and a great rule that the arts on reaching a peak should then decline, or perhaps it is part of man's nature to demand some license in creative expression as in other things, but we should not be surprised to see that ancient architects have done the very things which we sometimes criticize in buildings of our own times.[1]

It is no accident that, some 200 years later, Tafuri, while taking Piranesi as the beginning of the 20th century avant-garde, writes about the true nature of Piranesi's reconstruction in this way: "the archaeological mask of Piranesi's *Campo Marzio* fools no one: this is an experimental design and the city, therefore, remains an unknown."[2] In other words, contrast to Nolli's *Nuova Pianta di Roma*, what is unknown is the colossal non-place, a "negative utopia", that *Campo Marzio* brings about, and it is this unknown aspect of the city that makes Piranesi's project another "systematic criticism" of the concept of place, this time on the urban plane, calling into question the very concept of the city.

Interestingly, given the development of late 20th century cities, Piranesi's *Campo Marzio* no longer appears as an ominously foreshadowing "negative utopia", but a realistic urban condition that has left the traditional city and its spatial and formal integrity behind. Hence, for instance, Alex Krieger's parallel between Piranesi and present-day North American cities:

> If we look, say, at greater Toronto or Atlanta or the outskirts of historic Boston, or at the most infamous of edge cities, Tyson's Corner

1 Ibid, pp. 166-167.

2 Manfredo Tafuri, *Architecture and Utopia: Design and Capitalist Development* (Cambridge, Massachusetts and London: The MIT Press, 1973), p.15.

in northern Virginia, do we not see environments as boundless and multi-centred, as conspicuously wasteful, redundant, and eccentric, as "un-master-planned", and just potentially chaotic, as Piranesi's Rome ? [1]

Krieger's observation not only connects Piranesi's Rome directly to the urban reality of the 20th century, but also, however coincidently it might seem, reminds us of the fact that, as one of the few pre-20th century images included in Koolhaas' *S, M, L, XL*, Piranesi's *Campo Marzio* is adopted as a front page to the essay entitled "What Ever Happened to Urbanism?" This alone seems to evince a historic link and intellectual affinity between Piranesi and Koolhaas. Or does it?

Koolhaas and the "Paranoid-Critical Tourism" of the Generic City

In an essay on Surrealism, Hal Foster describes the experiences of André Breton as an assistant in a psychiatric clinic at Saint-Dizier in 1916. He treated a soldier who believed that the war was a fake, with the wounded made up cosmetically and the dead on loan from medical schools. The soldier intrigued the young Breton: here was a figure shocked into a paranoid alternative reality that was also somehow a critique of this reality.[2] Exactly to what extent this experience at Saint-Dizier impacted Breton's later life as a surrealist is impossible to know, but the primacy of the phenomenon of paranoia to Surrealism can never be overestimated. For not only Breton but other surrealists as well, from Giorgio de Chirico to Max Ernst, tried to capture and explore the anatomization of paranoia in their art. In particular, "an instrument of primary importance" called the "paranoid-critical method" was injected into Surrealism by Salvador Dalí.[3] As he put it himself, the method is "the spontaneous method of attaining knowledge based on the critical and systematic objectification of delirious associations and interpretations."[4] Paradoxically, "meant to subvert or question the

1 Alex Krieger, "Conflicting Ideals for the City," *GSD News* Winter/Spring 1995, p.16.

2 Hal Foster, *Compulsive Beauty* (Cambridge, Massachusetts and London: MIT Press, 1993), p.xii.

3 André Breton, *What is Surrealism? Selected Writings*, ed. Franklin Rosemont (New York: Monad, 1978), p.136.

4 Dawn Ades, *Dalí* (London: Thames and Hudson, 1982), pp.119-126.

'scientific' certainties of a 'reign of logic'"[1], it is the commitment to "evidence + paranoia" that is supposed to provide the possibility to negotiate its way out of the world of delirium, and thus enables the subject to creatively pass to the plane of reality through the discovery of "objective" significance in the irrational. In other words, while what is given is irrational, delirious, and insane, it is the paranoid-critical activity that transforms the delirium of interpretation from the pathological phenomenon of paranoia into an adventure of poetic discovery leading finally to a creative conquest of the irrational. Interestingly enough, "The Conquest of the Irrational" is exactly the title that Dalí used for his text in which the issue of "paranoid-critical method" is brought into play.

Koolhaas' proclivity to a paranoid-critical approach appeared as early as in the *Exodus*. Here, as Fritz Neumeyer noted, one is presented with a psychological terrain, a "Freudian tableau" of the bizarre condition of the metropolis. And by virtue of a surreal mode of perception and representation, Neumeyer continued, "the hunger for reality could find abundant, exotic nourishment and the desire for contradiction could discover sudden surfaces of friction to lay bare the secret poetic content of this reality."[2] It is, however, in *Delirious New York* that Koolhaas explicitly refers to Dalí's paranoid-critical method. For Koolhaas, paranoid-critical method is a sequence of two consecutive but discrete operations:

> 1. the synthetic reproduction of the paranoiac's way of seeing the world in a new light with its rich harvest of unsuspected correspondences, analogies and patterns; and
> 2. the compression of these gaseous speculations to a critical point where they achieve the density of fact: the critical part of the method consists of the fabrication of objectifying "souvenirs" of the paranoid tourism, of concrete evidence that brings the "discoveries" of those excursions back to the rest of mankind, ideally in forms as obvious and undeniable as snapshots.[3]

---

1 Dawn Ades, "Photography and the Surrealist Text", in Rosalind Krauss and Jane Livingston, *L'Amour Fou: Photography and Surrealism* (London: Arts Council of Great Britain, 1986), p.159.

2 Neumeyer, "OMA's Berlin", p.38 and p.46.

3 Rem Koolhaas, *Delirious New York* (New York: The Monacelli Press, 1994), p.238.

This interpretation of the paranoid-critical method gives rise to a surgical scalpel that Koolhaas uses along with his journeys around the world, from Berlin to New York and from Atlanta to Singapore, to cut — often on a somewhat subconscious level — at the "delirium" of the 20th century metropolis, and to turn on architectural concepts on a rigorously conscious level. Thus, while Coney Island is taken as the "incubator" of Manhattan and the "clitoral appendage" of New York, and the high-density urbanity, or the "Culture of Congestion", embedded in Manhattan and particularly in the Downtown Athletic Club is hardly to be understood as physical congestion alone. Rather, it is first of all a programmatic density that some years later triggers OMA's competition proposal for the Parc de la Villette in Paris where the "invisible congestion" is put onto work. In *Delirious New York*, the interpretation in the ethos of "paranoid-critical method" also takes the form of the "exaggerated extrapolation of an essentially unconscious Metropolitan landscape" known as *The City of Captive Globe*. "Devoted to the artificial conception and accelerated birth of theories, interpretations, mental constructions, proposals and their infliction on the world", Koolhaas wrote, *The City of the Captive Globe* is the capital of Ego, "where science, art, poetry and forms of madness compete under ideal conditions to invent, destroy and restore the world of phenomenal Reality."[1] Here, the "Culture of Congestion" of the metropolis is rendered into a rigid chaos in the form of the Manhattan grid in which each block represents an independent island with unique laws, a maximum agglomeration of different values — architecturally as well as ideologically, from Surrealism, Suprematism, Constructivism, Capitalism, to Le Corbusier's *Ville Radieuse* or the Berlin Wall...

The formation of OMA, the Office for Metropolitan Architecture in 1975 was announced by a poster where a tower building is hatched out of an egg reminiscent of the flower emerging from the egg in the *Metamorphosis of Narcissus* of Dalí. Meanwhile Koolhaas returned to Europe and to teach at the AA in London with certain dilemmas. As Koolhaas put it:

---

1 Ibid. p.294.

> USA: post-modernism triumphant. Europe: historicism on the rise — the "new" superseded, maybe forever? USA: freedom from context. Europe: context is everything. USA: everything big. Europe: everything small.[1]

This was accompanied by the establishment of OMA in Rottedam. Similar to OMA's Berlin, Rotterdam was once a historic center known for its own specific modernity between the wars. Heavily destroyed by the war, the city was rebuilt, but never regained its pre-war urban structure. Instead it became a model city of the mid-century urbanism because of its open center and above all the Lijnbaan, a linear shopping quarter designed by Bakema. Rotterdam later was considered as a "gigantic problem" where its open center was to be filled up with closed blocks in the name of urban renewal to make the city more "urban". In this sense Rotterdam epitomized the situation in Europe in the age of the "Reconstruction of the European City", as IBA claimed as its battle cry.

Koolhaas' view of Rotterdam at the time was quite different. For Koolhaas, not only did Rotterdam's unique quality depend on the openness of its center, but, as in the case of Berlin, its richness was inconceivable without the prototypical sequence of its mutation and destruction. Here, rather than the approach adopted by the "Reconstruction of the European City" that set out to erase the very fact of its history (the destruction of the world wars) in the name of history, Koolhaas turned his eyes to "the terrifying beauty of the 20th century". This is no easy task, for when order, taste and integrity are gone, "the resulting landscape needs the combined interpretative ability of Champoleon, Schliemann, Darwin and Freud to disentangle it."[2] What is at stake in this regard is, therefore, not so much an issue of taste or aesthetics in normal sense, but a way of thinking that means a further development of the "paranoid-critical method". Koolhaas writes:

> If there is a method in this work, it is a method of systematic idealization — a systematic overestimation of what exists, a bombardment

1 *S, M, L, XL*, p.518.

2 Ibid. p.206.

of speculation that invests even the most mediocre aspects with retroactive conceptual and ideological charge. To each bastard gets his own genealogical tree; the faintest hint of an idea is tracked with the obstinacy of a detective on a juicy case of adultery. [1]

For Koolhaas, however, the "paranoid-critical method" is more than a lens through which to examine the "delirium" of the 20th century city. It also offers him a surgical corallary to operate differently in the urban projects that he got involved, such as the project for the Boompjes Tower Slab in Rotterdam and the projects for Kochstrasse and Friedrichstrasse in Berlin, both done approximately at the same time. With these projects, postmodern "historicism", the first of the three dilemmas that Koolhaas faced in the late '70s, was overcome. Meanwhile, the second dilemma in terms of context and non-context remained unresolved. For even in the "terrifying beauty of the 20th century" where "each architectural doctrine contradicts and undoes the essence of the previous one as surely as day follows night", there is a context to explore in which, as we have seen, "each bastard gets his own genealogical tree; the faintest hint of an idea is tracked with the obstinacy of a detective on a juicy case of adultery."

To be sure, this context is no longer to be understood in the sense of figure-ground contextualism proclaimed above all by Colin Rowe. Nor does it make sense to label the clinical inventory according to the actual condition of each site as the Boompjes Tower Slab in Rotterdam or the projects for Kochstrasse and Friedrichstrasse in Berlin as urban contextualism. Yet "context" remains an architectural issue simply because of the fact that any building is located on a specific site, whose condition tends in one way or another exert impact on the building. Ultimately, as one can see in many of Koolhaas' projects from that time, it means an unreserved acceptance yet an imaginative approach to reality in all its messiness and unpredictability.

On the other hand, if any obligation to context is, in the end, a reformist position that there is no longer a need for territories that are still undetermined, free, conquerable only by new orders *ex*

---

1 Ibid. p.208.

nihilo,[1] then Koolhaas' position is surely never that of a reformist. In the heart of urban chaos he aspires to imagine nothingness, believing "where there is nothing, everything is possible. Where there is architecture, nothing (else) is possible."[2] The break point is Bigness. Beyond a certain scale, Koolhaas declares, architecture acquires the properties of Bigness which jettisons the "art" as well as the "morality" of architecture. Bigness transforms the organizational, structural and interior/exterior relationships of architecture. Bigness renders what is traditionally in the hands of architects or planners uncontrollable. Bigness discards urban contexts. Bigness breaks "with scale, with architectural composition, with tradition, with transparency, and with ethics — the final, most radical break: Bigness is no longer part of any urban tissue. It exists; at most, it coexists. Its subtext is *fuck* context."[3]

With such a manifesto of Bigness, what Koolhaas concerns is not merely with the size or scale of building projects ranging from the small one like Villa dall'Ava to the extra-large at Euralille. Beyond the dilemma of urban context and non-context as such, the urban implications of Bigness can only be comprehended in terms of the concept of the city itself. Again Koolhaas went to America to find possible answers. "Sometimes it is important to find what the city is — instead of what it was, or what is should be. This is what drove me to Atlanta — an intuition that the real city at the end of the 20th century could be found there...."[4] So begins Koolhaas' writing on Atlanta. For Koolhaas, like many American cities, Atlanta is more like a vast region than a city, though it had and in a sense still has a downtown. In the 1960s and '70s, while the downtown area of many other great American cities lay in a deep state of despair, downtown Atlanta was being revived and even thoroughly rebuilt. This process, Koolhaas maintains, was not urban renewal in the usual sense; it was "a virgin rebirth: a city of clone" characterised by an accelerating redevelopment thanks to the architect/developer John Portman. Started by rebuilding one block at a time to form an urban system where buildings are connected

1 Jacques Lucan, "Contextualism and Universality", *Lotus International* 74 (1992).

2 *S, M, L, XL*, p.199.

3 Ibid. pp.495-502.

4 Ibid. p.835.

together by bridges and skywalks, Portman invented something called "atrium", an architectural device which in essence is a huge glass-roofed interior space surrounded by hotel rooms, office cubicles, restaurants, and other facilities. Each atrium then is "a replica as inclusive as downtown itself, an *ersatz* downtown", while the original downtown "disintegrates into multiple downtowns, a cluster of autonomies."[1]

Consequently, what seemed to help rehabilitate and stabilize Atlanta's downtown actually accelerated its demise in two ways. First, the traditional concept of the street as an outdoor space no longer makes sense: even when the street space remains, the actual life of the city now occurs within the buildings. Second, as the downtown gets atomized, its autonomous particles could go anywhere:

> Now *all* is city, a new persuasiveness that includes landscape, park, industry, rust belt, parking lot, housing tract, single-family house, desert, air-port, beach, river, ski slope, even downtown.... Atlanta is not a city; it is a landscape.[2]

Atlanta's new urban — or rather, as Koolhaas puts it, "post-urban" — landscape is a case of Bigness *par excellence*. Koolhaas himself makes no secret of the implication of this case in OMA's intellectual history, maintaining that it was the shock of the Bigness of the New World(s) that made what was already implicit in *Delirious New York* explicit, especially against the background of Europe. Or, as Sanford Kwinter observed, nothing in Koolhaas' intellectual production better verifies this elaboration from New York to Atlanta:

> In Koolhaas, the concept of "America" has always loomed large. It has served not only enormous aesthetic ends, but has played a major role in generating both the novelty and the radicality in OMA's work (especially in the primarily European context with which they have dealt).... For Koolhaas, America, although deeply studied and assim-

1 Ibid. pp.839-843.

2 Ibid. p.850.

ilated into his work, has always strategically been kept at a "dangerous" — and therefore creative distance: it has been constituted and skilfully maintained as the necessarily ragged, mythical gateway to the destabilizing, novelty-introducing outside.[1]

Kwinter's observation is perfectly accurate insofar as Koolhaas' world journey is restricted to Europe and America. However, beyond the New World, the development of modern Asia provides another new territory (thus his use of the plural "New Worlds") that Koolhaas comes to explore. In *S, M, L, XL*, the outcome of this exploration takes form first of an essay called "Singapore Songlines: Portrait of a Portemkin Metropolis ... or Thirty Years of *Tabula Rasa*", and then "The Generic City".

In what sense is Singapore a test-bed of Bigness for Koolhaas? Certainly not in terms of its physical size or population. Rather, as the title of the writing indicates, the condition of Bigness is construed as the *tabula rasa* process of modernization. Historically, *tabula rasa*, or beginning anew from a clean slate, is not a new phenomenon of modernization. In Europe, Haussmann's boulevards, Le Corbusier's *Plan Voisin*, and Ludwig Hiberseimer's housing project for downtown Berlin, all exemplified the*tabula rasa* approach. This approach soon came to be regarded as the worst sin of modernism, a sin to be eradicated at all costs. Thus, as Koolhaas describes it, "the city of Zurich has found the most radical, expensive solution in reverting to a kind of reverse archaeology: layer and layer of new modernists — shopping centers, parking, banks, vaults, laboratories — are constructed underneath the center."[2] In this regard, Koolhaas once again stands out as an anomalous figure overtly claiming "*Tabula Rasa* Revisited" through a highly hypothetical project for the redevelopment of La Défense in Paris:

What would happen if, even in Europe especially in Europe— we declared every building in the entire zone that is older than 25 years worthless — null and void — or at least potentially removable ?[3]

1 Sanford Kwinter, "Flying the Bullet, or When Did the Future Begin?", ARCH + 132, June 1996, p.80.
2 *S, M, L, XL*, p.1249.
3 Ibid. 1105.

But in Southeast Asia, as cities leap from the nineteenth century straight into the twenty-first, *tabula rasa* is not a hypothetical approach but has been practiced in reality to a historically unprecedented extent. Is what we see happening in Asia an unexpected product of the Pandora's Box of modernization started in Europe, or will Asian modernization mature in its own right to transcend the dilemmas between tradition and modernization that seem to have been so painful for Europeans?

It is true that thirty years' *tabula rasa* has almost entirely erased Singapore's past where the original village is now but a wreckage floating in an ocean of hygienic newness. However, Koolhaas argues, the wholesale *tabula rasa* planning would have been impossible to implement in Singapore if it had not been supported by aspects of the local cultural heritage, in particular "a hard-core Confucian shamelessness" that, while imposing itself as an ultimate power of efficiency in "taking care" of its people and conflating "the dross of feudalism and the cream of democracy","the serious demands — for more and more freedoms — of modern culture" have been avoided or even suppressed.[1]

On the other hand, while thirty years of *tabula rasa* goes on, the hunt for identity and character has never ceased even in terms of physical environment. Hence the stylized "Asian" villages and Chinese gardens, the revivalist reconstruction of temples, tower buildings with "Chinese" roofs, the clip-on regionalism exercised on mediocre modern shopping malls. Consequently, cultural heritage and ethnic identity are in fact overwhelming rather than absent, though what remains of the past is no more than kitsch.

From Singapore there is only one step to *Generic City,* the synthetic reproduction and compression of urban facts or "souvenirs" into an urbanism that is "as obvious and undeniable as snapshots to the rest of mankind", as Koolhaas explicated regarding Dalì's "paranoid-critical method". Not surprisingly, the Generic City is construed as multi-racial and multi-cultural. Here, located in tropical climate, its business is clearly manifested in the form of downtown towers, while its urban life is concentrated to shopping centres.

1 Ibid. p.1017 and p.1039.

Taken together, the Generic City is formless or "free-style" assemblage of three elements — roads, buildings, and nature, while the cycle of interdependence between history and identity is completely severed:

> ... it is nothing but a reflection of present need and present ability. It is the city without history. It is big enough for everybody. It is easy. It does not need maintenance. If it gets too small it just expands. If it gets old it just self-destructs and renews. It is equally exciting — or unexciting — everywhere. It is "superficial" — like a Hollywood studio lot, it can produce a new identity every Monday morning.[1]

Perhaps, Ian Nairn is right in suggesting that after the disaster of modernism, identity, character or sense of place is "not a fine art extra, it is something that we cannot afford to do without."[2] Nowhere is this more evident than the Generic City in which as soon as a hint of identity is discovered, it is utilized to the maximum:

> If it is water-facing, then water-based symbols are distributed over its entire territory. If it is a port, then ships and cranes will appear far inland. If it is Asian, then "delicate" (sensual, inscrutable) women appear in elastic poses, suggesting (religious, sexual) submission everywhere. If it has a mountain, each brochure, menu, ticket, billboard will insist on the hill, as if nothing less than a seamless tautology will convince.[3]

What Koolhaas conceives as the paranoid-critical tourism of the Generic City is obviously a parody of modern man's desperate need for identity, character, and sense of place in an age of consumerism when everything is to the end subject to consumption. This parody also suggests that what Ricoeur calls the "lowest degree of creative culture" is a counter product of this consumer culture. But it should be noted that, in claiming "mankind as a whole is on the brink of a single world civilization representing at once

1 Ibid. p.1250.

2 Cited from Edward Relph, *Place and Placelessness* (London: Pion Limited, 1976), p.32.

3 *S, M, L, XL*, p.1263.

a gigantic progress for everyone and an overwhelming task of survival and adapting our cultural heritage to this new setting"[1], Paul Ricoeur also makes a strong case of the "tragic law of the creation of a culture" that necessarily runs counter to the "false consciousness" of identity. Creativity, Ricoeur maintains, eludes all definitions. Even scandals are necessary.[2] Nothing seems more compelling at this point than Koolhaas' own theoretical and architectural practice to which *S, M, L, XL* comes to testify.

Conclusion

The discourse on place and the city have been entangled with the predicament of the past and the present, and the way out of this tends, more often than not, to be predicted on the privilege of the past. In this context, a sharply different strategy has loomed large with undiminished eloquence in Koolhaas, which is now agglomerated into a single monumental work under the title of *S, M, L, XL*. I have attempted in this essay to track this strategy to Piranesi in line with Tafuri's seminal study of the avant-garde and architecture. At the first sight, Piranesi's archaeological reconstruction of the *Campo Marzio* brings us unmistakably to the past, yet as Tafuri has incisively disclosed, behind the "archaeological mask" is an experimental design that bears Piranesi's real intention not only to challenge the rules of architecture and cities, but to put the concept of place into severe criticism as well. In the same vein, the paranoid-critical tourism of contemporary cities around the world that Koolhaas carries out during the past decades and is finally rendered into a fictional version called the Generic City constitutes an "*archaeologue* (= archaeology with more interpretation) of the 20th century" that "needs unlimited plane tickets, not a shovel."[3] Simply dismissing questions of what places or cities were or should be, and thereby focusing on the present and looking into what places or cities actually are in a rapidly changing world, Koolhaas has launched a discourse in which the very concept of place and the

1 Paul Ricoeur, "Universal Civilization and National Cultures" in *History and Truth* (Evanston: Northwestern University Press, 1965), p.271.

2 Ibid, p.281.

3 *S, M, L, XL*, p.1263.

city is questioned, transgressed and finally subverted.

But there ends the parallel between Piranesi and Koolhaas. For while in Piranesi the city is subject to an experimental design, the epic tone of this experiment stems from, according to Tafuri, the struggle "between the demand for order and the will to formlessness"[1]. In other words, inasmuch as the city is a creative product of designers and planners, the will to design is of vital importance, whether in terms of longing for order or play of formlessness. This gave rise to, in the context of the early 20th century urbanism, two divergent and in fact what seemed categorically opposed approaches to urban conquest, epitomized by the urban ideal of Le Corbusier's *Ville Radieuse* on one hand and the avant-garde techniques of collage and montage on the other. Both incarnated, however, as Tafuri has judiciously pointed out, the modernist anxiety about the unprecedented urban upheaval unleashed since the Industrial Revolution.

In Koolhaas' paranoid-critical tourism of the Generic City, by contrast, the new conditions of architecture and urbanism in the late 20th century have not only changed the very concept of the city, but made urbanism as a profession in its traditional sense impracticable, and thus rendered the struggle between order and chaos meaningless. For Koolhaas, if our cities are formless rather than formally ordered, chaotic rather than geometrically structured, it is not because they are designed to be so but for the reason that they are the outcome of real forces, flows of capital in the first place, in operation of the late 20th century city. This brings us back to the essay entitled "What Ever Happened to Urbanism?" where the plate of Piranesi's *Campo Marzio* appears, for indeed no statement throughout *S, M, L, XL* is more capable of incarnating the disparity between Piranesi and Koolhaas than the question posed at the end of the essay: "What if we simply declare that there is no crisis — redefine our relationship with the city not as its makers but as its mere subjects, as its supporters?" Yet, as the essay concludes, "More than ever, the city is all we have."[2]

Thus, if the architectural and urban implication of Piranesi's

---

1 Tafuri, *Architecture and Utopia*, p.16.

2 *S, M, L, XL*, p.971.

"archaeological reconstruction" of *Campo Marzio* is to be understood against the background of an architectural ideology based on the Enlightenment concept of Reason and its critique, then what has been construed by Koolhaas urban discourse must be treated in relation to the epistemological break along with the rise and fall of modernism. But once again, it was through the examination of the historicity of the post-Piranesian avant-garde in relation to the capitalist development that Tafuri "effected a sharp epistemological break in contemporary history", a break that while "countering historical discourses content to build bland genealogies and consolidate the myth of the so-called 'Modern Movement', Tafuri instead revealed a set of subversive aporias that made it impossible to pursue recitals predicated on the combined expectations of the profession and the Academy."[1] Hence his critique of the ideological and utopian crisis of modern architecture. For Tafuri, this crisis, resulting in the demise of the architect as social ideologue, is only the final testimony of some "impotent and ineffectual myths, which so often serve as illusions that permit the survival of anachronistic 'hopes of design'."[2]

Whether or not sympathized with the political position of Tafuri, Koolhaas' strategy and discourse seem to have consolidated a post-Tafurian position in which Tafuri's assertive advocacy of an architecture that, after all, would not accept the terms of reality is finally discarded. Instead, he is determined to embrace the urban consequences of the 20th modernization and late-capitalist globalization, however relentless these consequences might be. He comes up with the idea of *Bigness* in a deliberate homage to the modernist strategy of *tabular rasa* yet without being overshadowed by the fear of chaos or the uncontrollable forces once embedded in the modernist penchant for order and aesthetic formalism. Faced with the new conditions, he is exalted rather than anguished or even traumatized, maintaining that unprecedented, unanticipated, and thus potentially liberating possibilities could revitalize an exhausted profession. To Koolhaas' mind, it is exactly because of the fact "it is out of control, the urban is about to become a major vector

1 Jean-Louis Cohen, "Ceci n'est pas une histoire", *Casabella* 619-620, January-February 1995, pp.48-53(49).

2 Tafuri. *Architecture and Utopia*, pp.181-182.

of the imagination. Redefined, urbanism will not only, or mostly, be a profession, but a way of thinking, an ideology: to accept what exists."[1] This implies, in essence, as Alejando Zaera has suggested, "a strategic retreat, the cessation of ideological resistance to the development of contemporary civilization."[2]

Needless to say, inasmuch as the evolution of human civilization is always flawed by its dark sides, and inasmuch as our understanding and handling of reality are necessarily value-laden, Koolhaas' strategy of non-resistance will in itself meet with opposition — in fact, from the moment of their appearance, most of Koolhaas' ideas have never failed to elicit controversy and outrage. Yet it remains to be a proof of Koolhaas' intellectual strength that, reading through the pages of *S, M, L, XL*, one feels more often than not that, to borrow Nick Land's words on Georges Bataille, "to accept his writings is an impossibility, to resist them an irrelevance. One is excited abnormally, appalled, but without refuge."[3]

---

1 *S, M, L, XL*, p.515.

2 Alejandro Zaera, "Notes for a Topographic Survey," *El Croquis* 53, *OMA / Rem Koolhaas 1987-1993*, March 1992, pp.32-51(50).

3 Nick Land, *The Thirst for Annihilation: Georges Bataille and Virulent Nihilism* (London: Routledge, 1992), p.xii.

of the manifestation. [illegible] will not only be [illegible] professional [illegible] way of thinking, [illegible] [illegible] of course, as A. [illegible] has suggested [illegible] the [illegible] of [illegible] to the development of contemporary mathematics.

[illegible] is always flawed by [illegible] and handling of reality [illegible] necessarily [illegible] of non-existence [illegible] itself [illegible] from the [illegible] of their [illegible] ideas have [illegible] to [illegible] Yet it [illegible] of [illegible] intellectual [illegible] that [illegible] through the pages of [illegible] one [illegible] than [illegible] that [illegible] we [illegible] the [illegible] but [illegible]

11

斯坦·艾伦《实践拒绝计划》导读[1]

1 最初发表于《时代建筑》2008 年第 2 期（总 100 期），收入本文集有修改。

我一直把斯坦 · 艾伦（Stan Allen）视为理论和实践兼顾的一类建筑师的代表之一。作为这样一位建筑师，他与那种只顾生产而没有思想的建筑师（这种建筑师包括两类，一类是忙于建筑产品生产的商业建筑师，另一类是忙于建筑作品生产的明星建筑师）或者艾伦在本文中称之为“唾弃理论”的“传统建筑实践”意义上的建筑师的区别自不待言，倒是艾伦对理论问题的思考更为发人深省。

斯坦·艾伦《实践：建筑，技术与再现》英文版（2000）

乍看起来，艾伦此文的出发点涉及的应该还是那个“批评”与“后批评”“批判性建筑”（critical architecture）与“投射性实践”（projective architecture）的问题。可以说，经过《时代建筑》“建筑历史理论”栏目的系列翻译和介绍，这些问题及其争论焦点的大致轮廓已经展现在我国读者的面前，关心这个栏目的读者也一定对相关的基本概念如“批判性”（criticality）、“投射性实践”（projective practice）、“设计智慧”（design intellegence）、“多普勒效应”（the Doppler Effect）等有了初步的、尽管在大多数情况下也许还是模糊不清的了解。艾伦的文章是这个系列的最后一篇，它是艾伦2000年出版的文集《实践：建筑、技能与再现》（*Practice: Architecture, Technique and Representation*）[1]一书的自序。

一般认为，艾伦与迈克尔 · 斯皮克斯（Michael Speaks）、罗伯特 · 索莫（Robert Somol）、莎拉 · 怀汀（Sara Whiting）等形成了一个“后批评”或者说“投射性”学派或者阵营。“后批评”的说法源自乔治 · 贝尔德的《“批评性”及其不满》（“Criticality and Its Discontents”）一文，[2]但是似乎并没有得到当事人的认可，甚至还遭到明确的反对；而“投射性建筑学”的提法则直接来自索莫和怀汀本人。[3]然而，无论冠以什么名称，我们应该有很多理由相信有这样一个“学派”或者“阵营”存在（或许我们可以暂且称之为“反批判性”阵营）。就本期介绍的艾伦而言，不仅他曾经在不同的场合直接或间接地引用斯皮克斯、索莫和怀汀的文字和观点，而且透过本期刊登的《实践拒绝计划》一文的字里行间，我们也不难看

1 Stan Allen, *Practice: Architecture, Technique and Representation* (Amsterdam: Overseas Publishers Association, 2000), p. XIII-XXV.

2 乔治 · 贝尔德：《“批判性”及其不满》（都铭译），《时代建筑》2007年第3期（总95期），第54-57页。

3 见罗伯特 · 索莫、萨拉 · 怀汀：《关于“多普勒效应”的笔记和现代主义的其他心境》（范凌译），《时代建筑》2007年第二期（总94期），第112-117页；以及范凌：《围绕〈关于“多普勒效应”的笔记和现代主义的其他心境〉一文的五点对话》，《时代建筑》2008第1期，第122-127页。

出艾伦在某些基本问题上与上述人物的相似之处。

在这篇自序中，艾伦一方面指出了旨在重复建筑学既有规范的“传统型实践”的不足，另一方面又着重对理论的地位和作用提出了挑战。在艾伦看来，建筑学是一种“物质性实践”（material practice），它充斥着偶然性和不一致性，然而理论却常常对它们视而不见，试图用规范的话语或者庞大叙事的计划将它们统一起来，而且这些话语或计划常常还是从别的什么地方（即非建筑学专业的）生搬硬套过来的。这样的观点不仅令人想起反“批判性”阵营对艾森曼的俄狄浦斯式的攻击，因为从乔姆斯基（Noam Chomsky）的语言学到德里达（Jacques Derrida）的解构主义，艾森曼的“批判建筑学”一直以移植和借用别的学科的话语和规范著称；或者对塔夫里（Manfredo Tafuri）“历史计划”（historic project）的不满，因为它在“建筑的意识形态批判”之中反而将复杂的历史情况简单化了；而且即使理论的话语来自建筑学本身（如艾伦在文中提及的类型学理论、建构理论或历史先例的理论），理论也常常因为急于建立规则而忘记建筑学作为“物质性实践”固有的多重性和细微的差异。显然，通过柯布的卡彭特视觉艺术中心和赖特的古根海姆美术馆的例子，艾伦试图说明的正是建构理论的不足，因为建筑师在实际操作的过程中思考的问题和为问题寻求的解答远比理论更复杂和细腻。同样的问题也在艾伦援引的米歇尔 · 德塞杜（Michel de Certeau）的城市步行者（the walker in the city）与空间系统中日常生活实践的不确定轨迹（the errant trajectories of everyday practices）的关系中得到进一步说明。

所有这些，包括艾伦文中提出的“实用的实践观”（pragmatics of practice）、“性能实践”（performative practice）、建筑学的“工具性”（instrumentality）的观点，或者他在文章伊始对艺术和文化评论学家戴夫 · 希基（Dave Hickey）的引述等都与斯皮克斯、索莫和怀汀等人不无相似之处，体现了对理论的质疑和不信任，它旨在寻求一种没有理论预设的、更加灵活和随机应变的、也更加关注具体操作性的建筑学。

但是，如果说共性确实存在的话，那么差异同样不可忽视。在艾伦的这篇文章中，既没有“后批评”也没有“投射性”的概念出现，取而代之的是对建筑学的“物质性实践”和“操作技能”（the mechanical techniques of practice）的强调。按照贝尔德的划分，对“批评性”的挑战是以斯皮克斯为其一，索莫和怀汀为其二在 2002 年分别发表的两篇文章为标志而形成，但是艾伦的这篇文章出现的时间显然更早。这是一个值得注意的细节，它再一次说明，如果建筑学思想会在某个时期出现共性或者“学派”的话，那么它的形成从来都不是评论家或史

学家常常乐于圈定的某个标志性事件在一夜之间发生的，其内涵也不是以一两个事件或术语就可以代表得了的。就此而言，个性也许比共性更值得关注，因为否则我们就会有重蹈"时代精神"的覆辙或者以一概全的危险。

因此，笔者更愿意在这里强调艾伦思想的独特性。在笔者看来，这种独特性就来自于艾伦理论和实践兼顾的双重身份的独特性。与斯皮克斯夸夸其谈"设计智慧"和"理论之后"的数字化设计之路的智者姿态不同（而且这种姿态还常常是以反理论的面貌出现的），艾伦走的是一条理论与实践兼顾的道路。诚然，无论在理论还是在实践方面，艾伦都远不能说是多产，而且已有的设计大多数都没有建成。但是，理论与实践并行的基本立场已经与斯皮克斯相去甚远。说到理论问题，尽管艾伦文中充满了对理论的质疑（这一点是斯皮克斯在《理论之后》中特别强调的），但是细心的读者也会发现（至少在本文中是这样），艾伦真正反对的是"诉求于计划"的理论，"一种指导性的理论构想"（an overarching theoretical construct），或者说"一种稳定的构造和衡量、测试个体行为是否与之吻合的持久不变的标尺"（a static construct, a persistent template of beliefs against which individual actions are compared, and tested for conformance）的理论。与之不同，艾伦将自己的理论工作称之为"写作"（writing），其目的"既不是为了解释，也不是为了证明某个特定设计工作或者工作方法的理由，而是一种澄清思想的连续过程"（not so much to explain, or to justify a particular work or working method ... as it is a continual process of clarification）。在艾伦看来，这样的理论写作是建筑学实践的组成部分，它与制图、建造或者教学并行，并且缺一不可。因此，艾伦的思想是具体的，强调建筑学的物质性和操作过程（场地操作、制图操作、写作操作、建造操作，等等）。

也正是在后面这一点上，我们可以看出艾伦与索莫和怀汀之间的区别。事实上，从斯皮克斯到索莫和怀汀，艾伦无疑是反"批判性"阵营中最全面且最具设计才华的一位。在笔者看来，这一点对于我们讨论的问题并非无足轻重，因为如果说"投射性"要求放弃传统意义上的"理论"与"实践"的对立，转而通过"设计研究"寻求理论与实践的融合的话，[1] 那么设计上的优势不仅为艾伦的"研究"工作奠定了充实的基础，也赋予艾伦的"理论"思维与索莫和怀汀不同的特质。索莫和怀汀从玄学意味十足的"多普勒'效应'"出发，为说明"投射性"一会儿求助于"绿色圆点"，一会儿又将自己的某个项目作为"投射性建筑"的典型。

1 范凌：《围绕〈关于"多普勒效应"的笔记和现代主义的其它心境〉一文的五点对话》，《时代建筑》2008 年第 1 期，第 122-127(124) 页。

为说明“批判模式”和“投射模式”的区别，认为艾森曼的圣地亚哥文化之城和库哈斯的西雅图公共图书馆可以引发具有实质意义的“哇！——什么？”（Wow! - What?）和“什么？——哇！”（What? - Wow!）之类的感叹的说法更是令人啼笑皆非，毫无说服力可言。相比之下，艾伦没有也无需这样做。

对于本文的翻译，project 是一个关键因而也值得斟酌的字眼。通常，与 project 对应的中文概念有“设计”“项目”“投射”和“计划”等。在笔者看来，如果把“设计”与“理论预设”的意义联系起来，用“设计”来翻译 project 也未尝不可，但是由于艾伦在本文中也使用了 design，再把 project 译成“设计”就必然会导致不必要的混淆。本文的 project 没有“项目”的意思，所以它应该在排除之列。“投射”的意思倒与索莫和怀汀所谓的“投射性”（projective）不无关系，但是，如果说索莫和怀汀的“投射性”与艾伦强调的“实践”有某些共同之处的话，那么艾伦在这里显然是把 practice 放在 project 的对立面，并且在二者之间旗帜鲜明地拥护前者而反对后者（这也就是为什么我们用“拒绝”来翻译本文题目中的“vs.”的原因）。就此而言，把艾伦的 project 译成“投射”即使不是错误的，也是不合适的。最后似乎就只剩下“计划”了。但是，用“计划”来翻译 project 并非纯粹是无奈之举。一方面，正如前面已经说过，“计划”令人直接想起埃森曼意义上的“批判计划”或塔夫里意义上的“历史计划”，而这些都是从斯皮克斯到索莫和怀汀所反对的，符合相关讨论的内涵。另一方面，艾伦所谓的 practice 与他援引的德塞杜思想中的“everyday practice”概念有着直接的关联，用它来对抗 project，就是反对“计划”的预设性、理念性、“宏大叙事”性和意识形态化。用索莫和怀汀的“多普勒效应”和“投射性”理论的术语来说，“计划”是“高定义”（high definition）的，是“热的”（hot），而“实践”则是“低定义”（low definition）的，是“冷的”（cool）。

但是，值得指出的是，如果我们因此如索莫和怀汀所说，将“热”视为“极度的艰难、繁琐、努力、复杂”，将“冷”视为“轻松悠然”，[1] 甚至进一步将“冷”延伸为随心所欲和玩世不恭的话，那么我们就有可能大大误解了艾伦和他的思想。在笔者看来，无论在写作实践和设计实践上，艾伦的工作一点也不“轻松悠然”。相反，他走的是一条艰难的多元实践的道路。之所以说是艰难的，不仅因为艾伦的实践包括设计、写作、制图、建造甚至教学，而要同时驾驭这些不同的实践模式显然绝非易事，而且更因为艾伦所谓的“物质性实践”既要有务实的建筑学专业技能和

1 罗伯特·索莫、萨拉·怀汀：《关于“多普勒效应”的笔记和现代主义的其他心境》，第 116 页。

艾伦　左：《点＋线——关于城市的图解与设计》中文版（2007）　右：《实践：建筑、技术与再现》英文版（2009）

对建筑学传统及其经典作品的充分理解和认识，又要有对当下建筑学操作条件的敏锐和把握以及由此发展而来的现实化概念基础，还要有高超的设计和建造能力，一种“通过创造新的物体和新的物质组织方式改变现实”的能力。就此而言，艾伦在本文中格外重视福柯所说的古希腊“技能”（techne）概念就不是什么奇怪的事情了。

因此，如果读者有兴趣深入了解艾伦在本文中阐述的思想，那么笔者认为，与其继续在索莫和怀汀的“多普勒效应”和“投射性”的概念中兜圈子，或者拘泥于贝尔德“批评”与“后批评”的划分，还不如去看艾伦以本文作序的《实践：建筑、技能与再现》一书中收集的文章，或者艾伦此前出版的另一本书《点＋线——关于城市的图解与设计》（*Points + Lines: Diagrams and Projects for the City*），该书已由建工出版社翻译出版，任浩译。顺便说一句，该书中文版中将“Field Conditions”译为《场所环境》值得商榷，因为它不仅没有体现原文的思想，而且在现有的建筑学中文语境中十分容易引起误解。

后记 2016

正如导读中已经提及的，艾伦这篇文章是他2000年出版的文集《实践：建筑、技能与再现》一书的自序。《时代建筑》于2008年译介这篇文章。次年，该书第二版问世，由英国拉特利奇出版社（Rutledge）出版，并且出现了一些有趣的变化，其中最大的变化是将第一版的“介质”（Media）一章替换为以“城市”（Cities）为标题的章节，并且在这一章中以《从物体到场域：建筑与城市中的场域条件》（“From Object to Field: Field Conditions in Architecture and Urbanism”）为题收录了《点＋线——关于城市的图解与设计》中的《场域条件》（“Field Conditions”）一文。

勒·柯布西耶　左：多米诺住宅　　右：哈佛大学卡彭特视觉艺术中心

就该书的“自序”而言，艾伦的修改主要集中于前半部分，大抵是将第一版中对理论与实践关系的阐述换了一些新的说法。在笔者看来，相较于这些修改，后半部分几乎没有修改的对 technique 的论述似乎更值得关注。因为正是在这些论述中，艾伦以更为建筑学的方式阐述了“实践”（practice）与“计划”（project）的不同。在这里，艾伦是通过两个具体案例做到这一点的。第一个案例是柯布为哈佛大学设计的卡彭特视觉艺术中心（The Carpenter Center for Visual Arts）。这个作品也是该书“建筑”（Buildings）一章的第三篇论文《勒·柯布西耶与现代主义动感：卡彭特视觉艺术中心》（“Le Corbusier and Modernist Movement: The Carpenter Center for Visual Arts”）论述的内容。作为柯布晚年的作品，卡彭特中心体现出与其早年机械美学时期作品的许多不同，如从多米诺体系加砖头填充墙体再加抹灰刷白到粗野混凝土（béton brut）结构和墙体的变化，或者从多米诺体系的方柱向卡彭特中心的圆柱的过渡，以及坡道等动感建筑元素的外在化等。平板无梁楼盖的具体做法也有发展和变化，从多米诺体系的下层空心砖和上层混凝土板的组合到卡彭特中心的下层混凝土板和上层空隙楼板（airfloor）的组合。但是总体而言，卡彭特中心的结构体系本质上仍然是多米诺体系的，它的一切变化都是为一个更为动感轻盈的空间塑造服务的。艾伦在序言中这样评述：

> 卡彭特中心以一种本质上并非轻型的材料取得了动感轻盈的感觉，这促使人们关注该建筑钢筋混凝土结构的某些技术层面的细节问题。它显示，创新和技术限制是与形式表现紧密联系在一起的。[1]

这令人想起柯林·罗（Colin Rowe）对柯布建筑思想与其建筑作品之间的关系的论述。一方面，罗将多米诺体系与雪铁龙住宅视为柯布建筑

1 斯坦·艾伦：《实践拒绝计划》（周凌译），《时代建筑》2008 年第 2 期（总 100 期），第 112-117(115) 页。

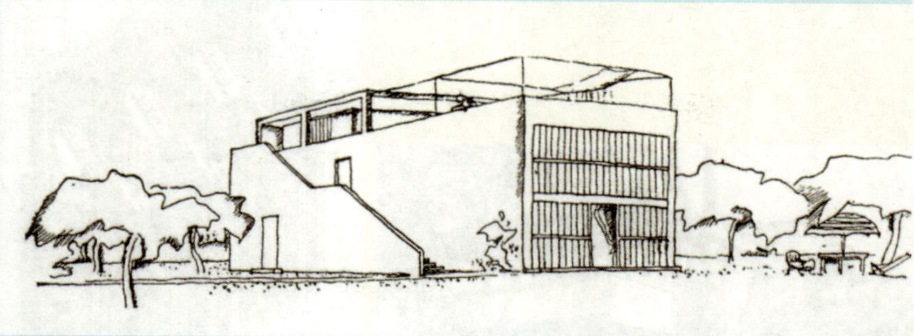

勒·柯布西耶　雪铁龙住宅

思想中两个最为重要的“概念”(concept)，前者是三明治式的，即板柱结构的垂直叠加，而后者则是美加仑（megaron）式的，即类似位于古希腊神庙核心位置的结构墙体围合。[1] 尤其是多米诺体系，罗在著名的《芝加哥框架》（“The Chicago Frame”）一文中，甚至将它称为“对建筑学这一普遍命题的回应”——“与其说它是一个结构，不如说它是一个偶像，一个确保原真的信物，一个新秩序的征兆，一个抵制个体放任的保证，一个能将动摇不定的表现主义纳入理性外观的规则。”[2] 但是另一方面，罗强调指出：

> 尽管对概念性实在的卓越分析一直是柯布的成就之一，他却很少在建成的作品中将分析当作解决方案来炫耀。他属于那种为数不多的、既重思想又重感性的建筑师。他总能在思想与感性之间维持一种平衡；……一方面，思想使感性进入某种文明状态；另一方面，文明性又借助于感性才得以实现。这一点是如此不言而喻，以至在柯布那里，概念性的主张从未真正成为设计的充分理由，而总是需要经过感知的重新过滤才能为设计所用。[3]

用艾伦“自序”的表述来说，多米诺体系就是柯布的建筑“理论”，一种能够成为“计划”（project）的“统领性的理论构想”。但是，柯布并没有使自己的实践沦为“计划”的奴仆，他通过柯林·罗所谓的“感知的过滤”使概念性的主张能够为设计所用。某种意义上，这就是艾伦旨在说明的“实践”与“计划”的不同之处吧。

如果说艾伦在论及柯布的卡彭特中心时并没有明确这样阐述，那么

1 Colin Rowe, “La Tourette,” *The Mathematics of the Ideal Villa and Other Essays* (Cambridge, Massachusetts and London, England: The MIT Press, 1976), p.197.

2 Colin Rowe, “The Chicago Frame,” *The Mathematics of the Ideal Villa and Other Essays*, p.107.

3 Colin Rowe, “La Tourette”, p.196.

赖特　左：葛尔住宅（1919）　右：全国人寿保险公司大厦设计方案（1924）

他通过另一个具体案例——赖特设计的纽约古根海姆博物馆——则相当明确地指出了“实践”与“计划”的这种差异。这也是“建筑”一章中的第二篇论文《重新认识古根海姆：所罗门·R. 古根海姆博物馆》（“Guggenheim Refigured: The Solomon R. Guggenheim Museum”）的要旨之一。在这里，赖特“有机建筑”（Organic Architecture）思想中关于结构与形式和空间的有机关系的主张被视为一种“理论”，或者是对于赖特来说一种能够成为“计划”（project）的“统领性的理论构想”。根据柯林·罗的“芝加哥框架”，从1909年的葛尔住宅（the Gale House）到1924年的全国人寿保险公司大厦（the National Life Insurance Skyscraper）设计方案，赖特都在致力于解决一个他深感棘手又欲罢不能的问题，这个问题来自于他对中性的、难以与空间和形式融为有机整体的《芝加哥框架》的不满，并最终导致他与自己的芝加哥前辈们的“分道扬镳”[1]。因此，艾伦指出，“赖特从来没有将结构理想化，在他的建筑中，结构永远不可能被抽象出来并自成一体。”[2]换言之，在赖特的建筑中并不存在多米诺体系那种普遍性的结构解决方案，他总是致力于采用与特定建筑的空间和形式“有机”

赖特　纽约古根海姆博物馆平面

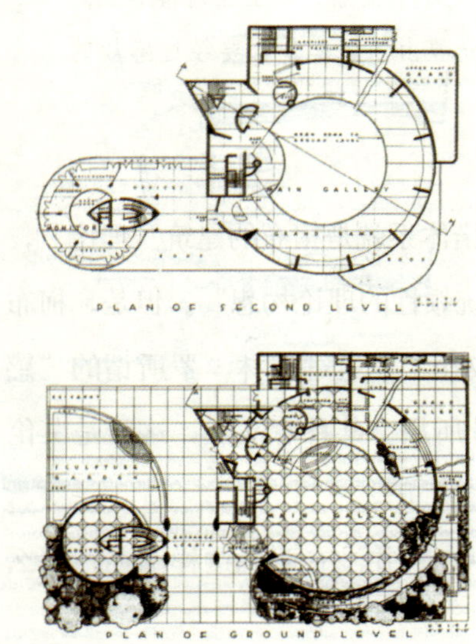

1 Colin Rowe, “The Chicago Frame”, p.93.

2 Stan Allen, *Practice: Architecture, Technique + Representation* (London: Rutledge, 2009), p.118.

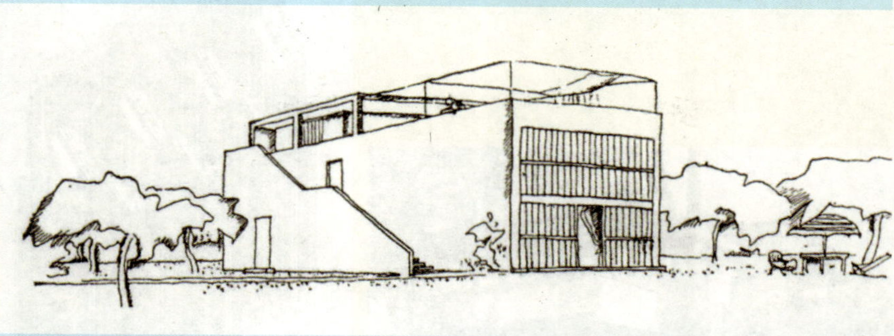

勒·柯布西耶　雪铁龙住宅

思想中两个最为重要的“概念”(concept)，前者是三明治式的，即板柱结构的垂直叠加，而后者则是美加仑（megaron）式的，即类似位于古希腊神庙核心位置的结构墙体围合。[1] 尤其是多米诺体系，罗在著名的《芝加哥框架》（“The Chicago Frame”）一文中，甚至将它称为“对建筑学这一普遍命题的回应”——“与其说它是一个结构，不如说它是一个偶像，一个确保原真的信物，一个新秩序的征兆，一个抵制个体放任的保证，一个能将动摇不定的表现主义纳入理性外观的规则。”[2] 但是另一方面，罗强调指出：

> 尽管对概念性实在的卓越分析一直是柯布的成就之一，他却很少在建成的作品中将分析当作解决方案来炫耀。他属于那种为数不多的、既重思想又重感性的建筑师。他总能在思想与感性之间维持一种平衡；……一方面，思想使感性进入某种文明状态；另一方面，文明性又借助于感性才得以实现。这一点是如此不言而喻，以至在柯布那里，概念性的主张从未真正成为设计的充分理由，而总是需要经过感知的重新过滤才能为设计所用。[3]

用艾伦“自序”的表述来说，多米诺体系就是柯布的建筑“理论”，一种能够成为“计划”（project）的“统领性的理论构想”。但是，柯布并没有使自己的实践沦为“计划”的奴仆，他通过柯林·罗所谓的“感知的过滤”使概念性的主张能够为设计所用。某种意义上，这就是艾伦旨在说明的“实践”与“计划”的不同之处吧。

如果说艾伦在论及柯布的卡彭特中心时并没有明确这样阐述，那么

1 Colin Rowe, “La Tourette,” *The Mathematics of the Ideal Villa and Other Essays* (Cambridge, Massachusetts and London, England: The MIT Press, 1976), p.197.

2 Colin Rowe, “The Chicago Frame,” *The Mathematics of the Ideal Villa and Other Essays*, p.107.

3 Colin Rowe, “La Tourette”, p.196.

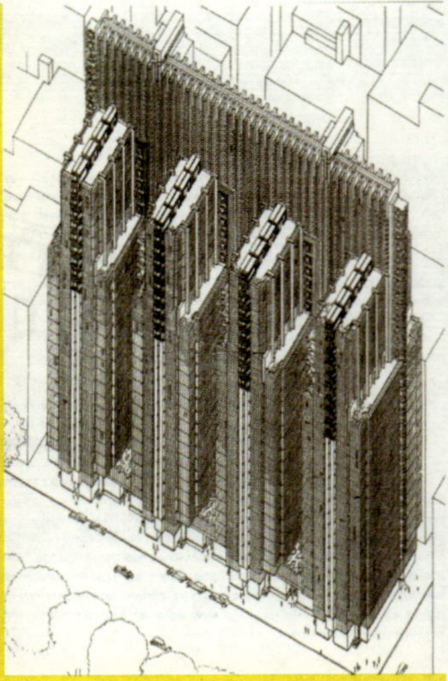

赖特　左：葛尔住宅（1919）　右：全国人寿保险公司大厦设计方案（1924）

他通过另一个具体案例——赖特设计的纽约古根海姆博物馆——则相当明确地指出了“实践”与“计划”的这种差异。这也是“建筑”一章中的第二篇论文《重新认识古根海姆：所罗门 · R. 古根海姆博物馆》（“Guggenheim Refigured: The Solomon R. Guggenheim Museum”）的要旨之一。在这里，赖特“有机建筑”（Organic Architecture）思想中关于结构与形式和空间的有机关系的主张被视为一种“理论”，或者是对于赖特来说一种能够成为“计划”（project）的“统领性的理论构想”。根据柯林 · 罗的“芝加哥框架”，从 1909 年的葛尔住宅（the Gale House）到 1924 年的全国人寿保险公司大厦（the National Life Insurance Skyscraper）设计方案，赖特都在致力于解决一个他深感棘手又欲罢不能的问题，这个问题来自于他对中性的、难以与空间和形式融为有机整体的《芝加哥框架》的不满，并最终导致他与自己的芝加哥前辈们的“分道扬镳”[1]。因此，艾伦指出，“赖特从来没有将结构理想化，在他的建筑中，结构永远不可能被抽象出来并自成一体。”[2] 换言之，在赖特的建筑中并不存在多米诺体系那种普遍性的结构解决方案，他总是致力于采用与特定建筑的空间和形式“有机”

赖特　纽约古根海姆博物馆平面

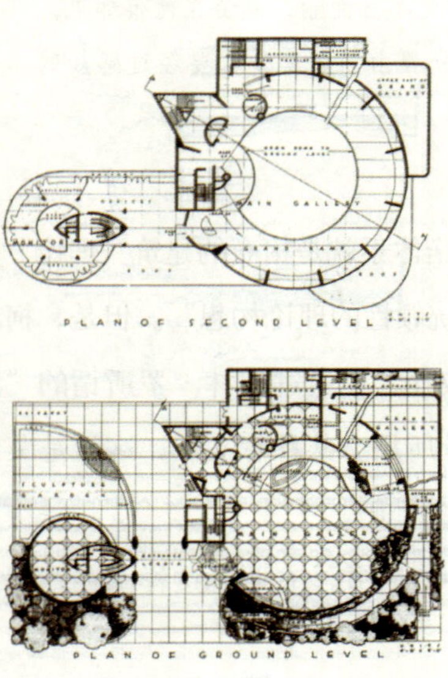

1 Colin Rowe, “The Chicago Frame”, p.93.

2 Stan Allen, *Practice: Architecture, Technique + Representation* (London: Rutledge, 2009), p.118.

结合的结构体系。

另一方面，即便是这个“有机”结合也未曾成为一个僵硬的教条，而总是审时度势，灵活掌握，甚至不乏妥协。就古根海姆博物馆而言，赖特的建筑方案旨在创造一个以螺旋坡道为载体的流动空间和形式，他最初的结构方案是在坡道之下设立柱子，但是很显然，这与赖特要求的有机连续性不相吻合。他随之走向另一极，设想了一个与整体的空间和形式关系完全一致的结构设想，将支撑结构与围合体系融为一体。但是，尽管该方案在建筑上有许多优点，从建造的观点看却未必切实可行。赖特最终接受的解决方案是，以圆形建筑平面的圆心为中心点，在靠近建筑外围的部分增加一系列具有支撑作用的放射形墙体。这个解决方案虽然在本质上与建筑的概念整体并不一致，但是却显示了自身的逻辑性和有效性。艾伦写道：

> 这一点不仅说明赖特灵活的策略以及他对建筑技术的直接把握，而且也揭示了尽管该建筑最终结果其实并不那么有机，但是它给人的体验还是一个整体和连续的结构。事实上，赖特并没有因为结构上的权宜之计而牺牲他所希望的连续性。[1]

就此而言，赖特古根海姆美术馆的设计过程不仅意义非凡，而且对于艾伦所论述的实践观念来说至关重要。简言之，艾伦指出，“赖特之所以能够运用多重结构原则有效地自由操作，恰恰是因为他将结构理性视为实践而非计划（structural rationality as practice, not as project）”。[2]

在艾伦看来，这是一种实用主义观念。它要求实施和行为的具体差异（concrete differences of performance and behavior），而不是理念和话语之间的抽象关系（abstract relations between ideas and discourses）。艾伦进一步写道：

> 对赖特和其他大多数让我感兴趣的建筑师来说，每一个建筑都是一次单独的初始尝试，而不是某个更大计划（a larger project）的组成部分。建筑物不是表达外来原则的样板，也不是测试理论定律的案例。特殊的情况要求特殊的解答。连贯性与合理性来自结构的硬性逻辑与惰性的材料本身。在赖特的例子里，结构理性并不是一个需要通过材料进行表现的绝对事实，而是一种机会和资源——一种需要自由处理的短暂的稳定性。在建造问题上，赖特的不同凡响之处在于，他知道应该在什么地方和什么时候进行妥协，知道应该如何看待整体性和“材料

1 艾伦：《实践拒绝计划》，第 116 页。
2 同上，第 115 页。

的真实性”的神话。赖特体现的是一种工作方法，它认为，建筑学创造感知经验和感官素材的能力——也就是说实际的结果和效应——远比顺应抽象的理论标准更为重要。[1]

没有什么能够比这更明确也更好地说明艾伦“实践拒绝计划”的含义了。有了这两个具体案例，艾伦在“自序”前半段对“实践”“理论”“计划”及其相互关系的种种较为抽象的表述就变得不是那么晦涩难懂甚至“惊世骇俗”了。简言之，“实践拒绝计划”并非“实践拒绝理论”。“实用主义”这个在“批评”与“后批评”或者说“批判性建筑学”与“投射性建筑学”的争论之后的建筑学主流立场——事实上，《哈佛大学设计杂志读本》（*A Harvard Design Magazine Reader*）就将《新的建筑实用主义》（*The New Architectural Pragmatism*）作为以这场争论为主题的第 5 期的标题[2]——也不是斯皮克斯所谓的“理论之后”。[3]用艾伦在“自序”第二版中引述的美国实用主义哲学家威廉 · 詹姆斯（William James）的话来说，“实用主义使理论变得灵活；它是理论的柔化，让理论发挥作用”。（Pragmatism unstiffens all our theories, it limbers them up and sets each one at work.）[4]

哈佛大学设计杂志读本《建筑与新实用主义》

1 艾伦：《实践拒绝计划》，第 116 页。

2 见 *The New Architectural Pragmatism : A Harvard Design Magazine Reader* | 5, ed. William S. Saunders (Minneapolis: the University of Minnesota Press, 2007).

3 迈克尔 · 斯皮克斯：《理论之后：建筑学院里关于理论价值及其对设计创新作用的争论》（周鸣浩译），《时代建筑》2007 年第 5 期（总 97 期），第 72-73 页。

4 Allen, *Practice: Architecture, Technique + Representation*, p.XIII.

12

中国建筑传媒奖之我见

及对《被公民的中国建筑与被传媒的中国建筑奖》一文的几点回应[1]

1 最初发表于《Domus 国际中文版》2011 年第 3 期（总 51 期），也见《走向公民建筑：中国建筑传媒奖／中国建筑思想论坛 2007—2010》，南方都市报 中国建筑传媒奖组委会 中国建筑思想论坛组委会编，桂林·广西师范大学出版社，2012 年，第 473-481 页。

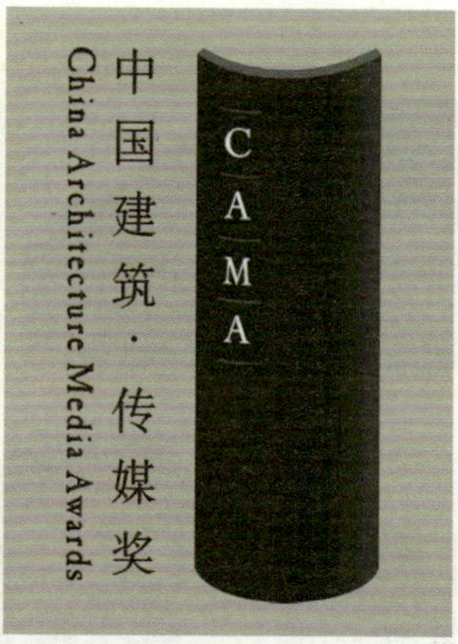

上：中国建筑传媒奖标志

下：中国建筑传媒奖《走向公民建筑》

由《南方都市报》《南都周刊》《风尚周报》等大众媒体发起并组织的“中国建筑传媒奖”已经举办两届。自 2008 年开始以来，这项活动得到了国内主要建筑媒体的积极响应和参与，成为中国建筑界一个不大不小的事件。说它不大，是因为尽管有国内主要建筑媒体（特别是其主编们）的鼎力相助，但是真正对之着力进行报道的还是它的主办媒体，其影响力自然有限；说它不小，是因为至少就第二届的提名名单而言，其阵容已不可小视，给奖项组织者和评选者的压力剧增，评奖结果自然也会引来种种争议。但是，如果我们希望把这个奖项继续搞下去，而且搞得更好，成为推动中国建筑和社会发展的积极力量，批评和争议是必需的。对于这个奖项的意义何在，具体的奖项应该如何设立，评选应该如何进行，其标准应该如何建立，当然，更重要的是，何谓“公民建筑”、“中国建筑传媒奖”能在中国公民社会的建设中发挥怎样的作用，这些问题都应该得到广泛的讨论。作为两届“中国建筑传媒奖”的终评会成员，我愿意借《Domus 国际中文版》一方空间谈谈自己在一些基本问题上的看法，同时对清华大学周榕老师的“被公民的中国建筑与被传媒的中国建筑奖”（以下称“被”文）做几点回应。

一　对第二届“最佳建筑奖”空缺的说明

本届“中国建筑传媒奖”之后，争议最多的也许是“最佳建筑奖”的空缺——至少笔者了解的情况如此。大凡评奖，出现奖项空缺并非什么空前绝后、足以令人大惊小怪的事情。它无非说明，评委根据他们理解的奖项标准和把握的尺度做出的某种抉择可能过严，就像评奖有时可能会过松一样；可能不当，就像获奖名单本身也会引发争议一样。争议不可避免，众口难调的情况总会出现。但是，“被”文的质疑非同一般，它以“最佳建筑奖”空缺为由，进而对整个“中国建筑传媒奖”发难。因此，在我们对“中国建筑传媒奖”进行任何可能的讨论以及对“被”文进行任何可能的回应之前，有必要对本届“最佳建筑奖”空缺这一结果的产生过程做一个基本说明。

左上：台湾宜兰津梅栈道　中上：香港钻石山火葬场　右：天津西青区张家窝镇小学　下：西安大唐西市博物馆

本届“最佳建筑奖”的入围作品有四个：台湾宜兰津梅栈道、香港钻石山火葬场、天津西青区张家窝镇小学、西安大唐西市博物馆，它们由提名和初评委员会经过数轮投票根据末位淘汰制产生。这个结果可能很好，也有可能不那么好，比如完全有可能在过程中淘汰了更好的作品。据本届提名和初评委员会主席赵辰老师事后介绍，尽管包括他本人在内的部分提名和初评委员会成员对这个结果并不十分满意，但是根据评奖章程，任何逆向操作（比如对已经淘汰的作品重新投票）都是不允许的。换言之，这一制度的设计是为了排除人为操纵的可能，但也完全有可能导致一种无法预料和不尽如人意的结果（事实上，笔者也认为如果有其他作品入围会更好），但这也是它为恪守投票章程而付出的代价。当然，人们也可以思考“中国建筑传媒奖”的章程是否有改进的可能，比如是否可以允许提名和初评委员会对淘汰作品重新投票而又避免腐败和暗箱操作的可能，不过这些都是“中国建筑传媒奖”今后在评奖制度的设置上应该考虑的问题。

按照组委会的设想，终评会应该在四个作品中评出“最佳建筑奖”的获得者。作为终评会成员之一，笔者本人根据评审材料做出的最初判断是比较倾向于台湾宜兰津梅栈道或天津西青区张家窝镇小学的。但是，有一个问题还不得不说：“中国建筑传媒奖”对“最佳建筑奖”和“居住建筑特别奖”设立了实地考察制度。因此，本届终评会将“最佳建筑奖”的评审放到最后，就是希望评委不仅对入围作品的评审材料进行消化和理解，而且有足够时间听取实地考察评委的考察意见，并在此基础上进行充分的讨论。值得强调的是，正是实地考察表明，尽管台湾宜兰津梅栈道、香港钻石山火葬场、天津西青区张家窝镇小学、西安大唐西市博物馆等四个入围作品都在各自不同的方面做出了可贵的努力和尝试，取得了值得肯定的成就，但也存在着种种较为严重的不足，难以达到“最佳建筑奖”获奖作品应该达到的标准。这直接导致了七位评委中有五人放弃投票。鉴于弃权评委过半，根据“中国建筑传媒奖”的评奖章程，本届“最佳建筑奖”最终空缺。

让我们看一看终评会对四个入围作品的主要意见（需要指出，以下意见是笔者个人的整理，而非终评会的正式陈述。事实上，这样的陈述在两届“中国建筑传媒奖”中都不存在，这也是今后需要改进的地方）：

台湾宜兰津梅栈道

该作品利用原有的机动车行桥梁，以巧妙而简单的构造，营造出步行栈道的空间，体现出建筑师因地制宜的设计理念，其结构和材料处理以及形式感方面也有诸多可称道之处。但是，该建筑在实际使用中出现不少问题，遭到使用者的抱怨甚至强烈不满，这也是该建筑师之前完成的部分作品的共性之一。

香港钻石山火葬场

新的钻石山火葬场提供了一个平静、祥和及庄严的环境和空间氛围，完成度也较高。但是评委会认为，该建筑在设计理念和空间创造方面还是过于平淡，未能在火葬场这类极具精神性和对生死意义进行终极思考的建筑设计上形成突破和新的探索。

天津西青区张家窝镇小学

该建筑在设计理念和体量关系上有许多值得肯定的地方，表现了建筑师对儿童教育的思考和关切，也具备较高的完成度。遗憾的是，作为“最佳建筑奖”的入围作品，评委会认为它在室外活动空间、食堂的空间处理、教工用房的使用功能与造型处理的关系等方面还存在诸多不足，甚至因局部结构的缺陷影响了走廊的通行高度。

西安大唐西市博物馆

在一片“新唐风”建筑群中，该建筑个性突出却不失与环境的对话与协调，设计师将古城的里坊空间布局特点贯彻于博物馆空间的设计之中。可惜的是，由于种种原因，该建筑在实施中遭遇困难，最终结果不尽如人意，中大厅空间的处理也比较令人失望，大厅玻璃顶的做法在保温隔热和日照光线与展示空间的关系方面亦遭到众多评委的质疑。

值得再次强调，评委们在评奖过程中关注到的上述问题在评审材料中常常难以体现，而只有在实地考察时才能有所感受，甚至强烈地感受。这也再一次说明实地考察应该是建筑评奖中十分重要的一环，而“中国建筑传媒奖”是目前国内众多建筑奖中唯一实行评委实地考察制度的奖项。在笔者看来，这也是“中国建筑传媒奖”另一个值得肯定的方面。

颁奖仪式上，主持人宣读了笔者与本届提名和初评委员会主席赵辰老师代表终评会起草的“最佳建筑奖”空缺说明：“‘最佳建筑奖’是‘中国建筑传媒奖’中最为重要、代表大奖核心价值观的奖项。它既要体现公民建筑的理念，最大程度彰显公众利益，还要在建筑的设计理念、完成度、使用感受等多方面表现出较高的水准。因而经过评委会充分讨论、慎重权衡，听取了考察评委的实地考察报告后，决定让本届的‘最佳建筑奖’空缺，虽然十分遗憾，却是为了努力维护大奖的高水准和严肃性，并衷心期待下届‘中国建筑传媒奖’能有更高水准的、更为令人信服的、最佳建筑奖’入围作品出现。”

换言之，本届“最佳建筑奖”的空缺有评奖制度上的原因，也有评委“求全责备”的原因，但“宁缺毋滥”、维护“最佳建筑奖”应该具备的高标准是多数评委在投票时的基本立场。笔者衷心希望，这一基本立场能够让“中国建筑传媒奖”在今后的发展中避免重蹈国内各类奖项滥、乱、烂而陷入信任危机的覆辙。

需要看到，尽管“最佳建筑奖”空缺，本届“中国建筑传媒奖”从成功和不成功的方面都进一步明确了自己的主题和奖项性质。它表明，本届“中国建筑传媒奖”终评会愿意从社会意义（用“被”文的术语来说就是“伦理价值”）和建筑品质（亦即“被”文所谓的“学术价值”）两个方面同时评判建筑的优劣。当然，如果一定要在二者之间做出某种偏重，那么它也认识到，对于“走向公民建筑”的宗旨而言，“社会意义”无疑具有不可取代的优先地位。正如本届终评会成员夏铸九老师的实地考察报告中指出的，如果建筑师的空间形式追求需要以牺牲实际使用或公共利益为代价，那么其作品或许可以获得建筑创意奖，却不是公民建筑奖应该期待的。夏铸九老师的报告还指出，“在台湾和香港，市民社会已经浮现，正有待进一步生长，中国大陆，市民社会刚刚浮现。而建筑，作为营造劳动的社会与历史分工，不只是在基地上放置一个容器而已。使用者与地方居民有发言和参与的权利，这是一种公共领域，一种公共空间的再现，它可以改变封闭的建筑论述和建筑师的惯行，有助于公民建筑的形成，它就是公民社会。”应该说，这正是本届终评会在“最佳建筑奖”的评审以及“居住建筑特别奖”“青年建筑师奖”“组委会特别奖”等其他奖项的评审中努力贯彻的标准。

二 “被公民”与“被传媒”——两个值得剖析的“被”字句时代的新说辞

不知从什么时候开始，中国社会似乎进入了“被”字句时代：“被代表”、“被和谐”“被增长”“被动车”“被高速”“被自愿”“被通过”“被小康”“被开心”“被寂寞”甚至“被自杀”……真可谓应有尽有，百试不爽。有理由认为，“被”字句时代折射的问题其实早已有之，即长期以来中国民众在影响和决定社会进程以及与己相关的日常生活方面的权利缺失。但是，“被”字句的出现和广泛流传却是近几年的事。应该说，它是中国社会进步的一种表现，因为虽然“被迫无奈”仍是“被”字句之所以能够借助互联网如此流行的社会根源，但是“被”字句得到如此广泛认可和运用本身就是“被”者自主性增强的表现，是某种意义上的公民意识的觉醒，因而也是值得肯定的。

然而，如同一切互联网流行语一样，人们在对“被”字句活学活用甚至滥用之时，也会使一个原本不无意义的表达方式转化为某种人云亦云、盲目跟风的行为模式，自以为与时俱进、玩文弄潮，实则花拳绣腿，是思想贫瘠的牺牲品。

“被”文以“被公民的中国建筑与被传媒的中国建筑奖”为题，其基本立论有二：“被公民”与“被传媒”。单以这一点来看，“被”文不可不谓打上了强烈的“被”字句时代的烙印。那么，在“被”文中，这两个基本立论究竟是怎样以“被”字句来讨论“中国建筑传媒奖”的问题的呢？

先说“被传媒”。一个由媒体举办的奖项被这个媒体以某种方式热烈报道，这原本符合事情的逻辑，也无可厚非。但是，读罢“被”文，笔者一个无法回避的感觉是，它似乎要以第二届“最佳建筑奖”空缺为由，将其说成是主办媒体为谋取自身的最大利益刻意策划的一次“被传媒”事件：“全程模仿奥斯卡颁奖礼的第二届中国建筑传媒奖颁奖礼终于被一个好莱坞式的戏剧性情节推向了高潮：最佳建筑奖空缺。这意味着，在2009和2010整整两年间中国落成的海量建筑中，竟没有一个能被评委会认可为‘公民建筑’的典范，从这个意义上说，中国建筑集体失败了。”乍看起来，这是对颁奖仪式过分“奥斯卡化”的指责，或是对“最佳建筑奖”空缺而令“中国建筑集体失败”的不满。然而，“被”文很快显示，它的真正论点在于，无论有无“最佳建筑奖”产生，“中国建筑传媒奖”其实都注定是一出“被传媒”的闹剧：“选出最佳建筑，自然可以顺理成章地推出一位平民化、亲民化的建筑英雄或建筑明星；选不出最佳建筑，就立刻成为一个颇具新闻点的爆冷事件。”因此，“被”文的结论只能是一个：“建筑输了，传媒赢了。”

笔者希望，本文前述对本届“最佳建筑奖”空缺的说明已经能够澄清，人为制造“被传媒”爆冷事件的可能和意愿都不存在——当然，面对“最佳建筑奖”空缺这样一个意想不到的结果，主办媒体将它说成“爆冷”则是另一回事。笔者也希望，这个说明能够澄清，本届“中国建筑传媒奖”对建筑伦理问题给予了特别的关注——只是，这种关注与“被”文崇尚的“不作恶”“不帮凶”的所谓伦理“底限”迥然不同（这个问题本文后面还将讨论）。笔者更希望，这个说明还能够澄清，在本届“中国建筑传媒奖”中，建筑没有输，因为它在关注建筑的“伦理价值”的同时，也对建筑自身的品质（亦即“被”文所谓的“学术价值”）予以了充分的重视——相比之下，倒是“被”文所断言的“在超速城市化进程中卖力出演的绝大多数中国当代建筑师都沾染了‘原罪’”着实为当代中国建筑蒙上了一层彻头彻尾的“集体失败”的色彩。

至于“被”文的“被传媒”造“星”说，笔者相信，过去的两届“中国建筑传媒奖”已经表明，获得其中任何奖项都无益于获奖者成为“被”文意指的那种“明星”。要么他或她已经是某种意义上的“明星”建筑师（如获得第一届“青年建筑师奖”的“标准营造”），要么他或她该干什么还是干什么（如获得第二届“居住建筑特别奖”的何勍、曲雷）。或许，“被传媒”造“星”说透露的不是别的，正是“被”文自己已经习以为常的“明星”式思维罢了。

有趣的是，尽管“被”文伊始就提出“被传媒的建筑奖”的立论，但是它似乎并没有完全站在“被传媒”的对立面；相反，在对“建筑输了，传媒赢了”的“被传媒”现象一番责难之后，“被”文又将矛头对准“中国建筑传媒奖”评委会中太多专业人士、太少大众传媒的人员构成。或许，把这个问题作为“被传媒”的一个方面进行讨论并不合适，因为按照“被”文的“被”字句逻辑，这个奖项应该不是“被传媒”而是“被专业”了，但是无论如何，在“被”文看来，“‘公民建筑’，本应是大众传媒对中国建筑界所进行的一次伦理学拷问”，可惜由于专业传媒的过分介入和过多专人人士的评委会构成以及过于专业的评判标准而“最终走过了场”。“被”文认为，大众传媒与建筑专业传媒的本质区别就在于“前者更关注建筑的社会伦理价值，后者更关注建筑的专业学术价值，两个媒体平台各自秉持的价值观虽不算泾渭分明，但侧重点明显不同”。

应该说，上述区分基本正确。问题在于，在批评发达的国家，即便大众传媒的建筑批评常常也需要这些传媒自身具有一定专业素养、甚至很高专业素养的评论家来完成。相比之下，这样的评论家在目前中国的大众媒体内基本不存在，这大概就是作为大众媒体的“中国建筑传媒奖”主办方愿意与专业媒体合作、并希望终审评委基本由专业人士组成的原因吧。事实上，无论提名还是初评和终评，“中国建筑传媒奖”都是一个由大众传媒搭台、专业人士唱戏的奖项，但是与其他专业奖项不同，它提出了一个特别的主题 ——“走向公民建筑”，其特别之处就在于，它着重“那些关心民生，如居住、社区、环境、公共空间等问题，在设计中体现公共利益、倾听人文关怀、并积极为现时代状况探索高质量文化表现的建筑作品”（引自“中国建筑传媒奖”《宣言》）。对于历来较少关注此类问题的中国专业建筑传媒和建筑学专业来说，这样的主题不仅十分需要，而且正如本文前述希望表明的，也在一个大体上由专业人士构成的评奖过程中得到基本贯彻。这一点已经无需赘述，可能的分歧应该在于，“被”文似乎认为，一个与建筑相关的大众传媒奖项只进行（或基本只进行）伦理学判断而无需（或基本无需）顾及建筑本身的品质，只要拉高调门轰轰烈烈一番即可；而在笔者以及本届终评会成员看来，只要是建筑奖项，无论是大众传媒的还是专业传媒的，在伦理问题和建筑自身的品质之间做出权衡和判断就无可回避。一个令人信服的“最佳建筑奖”难道不需要经过这样的权衡和判断就可以产生吗？获得第一届“中国建筑传媒奖”之“最佳建筑奖”的毛寺生态实验小学已经很好地说明这一点。

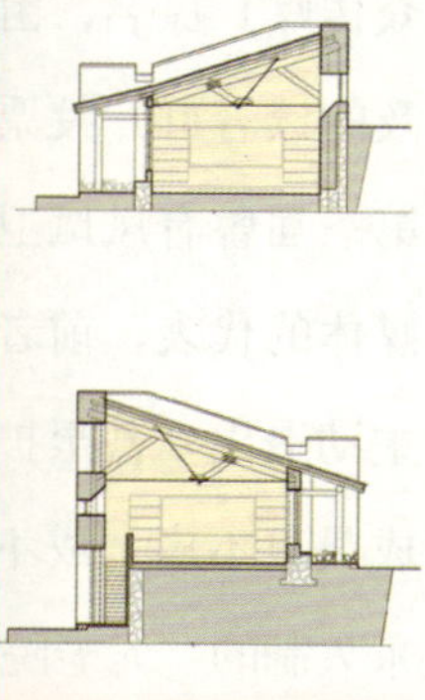

毛寺生态实验小学

相较于“被传媒”说，“被”文的“被公民”说也许更符合“被”字句时代的逻辑，在立论上也更为简单。在它看来，中国仍与公民社会相去甚远，而在这样的情况下提出“公民建筑”无异于“缘木求鱼”，使建筑“被公民”。在此，笔者无意对这样的观点是否可以在“被”文援引的列斐伏尔的“空间生产”理论中找到支持展开论证，也无意重复“被”文在这个问题上看似锋芒犀利实则粗糙混乱的论断。在有限的篇幅空间之内，本文只想提醒读者注意，“中国建筑传媒奖”的主题是“走向公民建筑”。宛如柯布的《走向新建筑》（更准确地说应该是《走向一种建筑》[1]），它表达的是一种愿望，一种应该为之努力的目标。毋庸讳言，中国社会的确还与“公民社会”相去甚远，但这并不表示我们不应该或者不能够通过自己的努力去推动公民社会的建设——如果我们认为公民社会是一个值得争取的目标的话，而“中国建筑传媒奖”就不可不谓其主办方和参与者为建设中国公民社会而做出的一种努力。也许，这样的努力诚如“被”文所言“很傻很天真”，但是笔者相信，相较于“被”文最终只能求助于“不作恶、不帮凶”的所谓伦理“底限”，这是一种更为积极和建设性的态度。

应该看到，对于“中国建筑传媒奖”而言，“被”文的“被公民”说也并非全无道理。“中国建筑传媒奖”以“走向公民建筑”为宗旨，却只有专业人士参加提名和评审，不仅如“被”文批评的那样大众媒体参与甚少，而且完全听不到公众的声音。诚然，“中国建筑传媒奖”的章程中已经赋予网上投票一定的地位（即网上得票最多者可以在终评中得一票），但是至少从已经举办的两届来看，网上投票者多为建筑专业的学生或者与某个入围者相关的专业人士，而非专业圈之外的社会民众。诚然，“走向公民建筑”的奖项未必只有社会公众的直接投票才能产生，但是完全没有社会公众参与的“走向公民建筑”奖项必然会削弱它的社会意义和影响力。在此，笔者建议下一届“中国建筑传媒奖”至少可以增设一个奖项，这个奖项完全由大众传媒（至少是作为主办方的三个大众传媒）提名，由社会公众（至少是作为主办方的三个大众传媒可以企及的读者群）投票产生，同时在终评会中恢复（之所以说恢复，是因为第一届曾有凤凰卫视节目主持人梁文道和南方都市报总编曹轲作为大众媒体的代表，前者由于遭受“政治封杀”而缺席本届终评会，后者则是主动退出）甚至扩大大众媒体的评委成员构成，而且即便这样，终评会成员也不应一成不变，用崔愷评委的话来说，终评会不应成为一个实行永久制的“元老院”。

1 关于这个问题，参见本文集《勒·柯布西埃 *Vers une architecture* 译名考》。

冯纪忠先生获第二届《中国建筑传媒奖》终身成就奖现场

三　当代中国建筑需要超越“不作恶”“不帮凶”的伦理“底限”

可以说，用力过猛而在言论上有失公允，这是“被”文的基本特征。尽管如此，“被”文还是有保留地表达了对“中国建筑传媒奖”的肯定：“尽管，‘公民建筑’的评选在今日之中国还是缘木求鱼，但至少这样的活动或许有助于中国建筑界进行有关建筑伦理的些微反省。”毋庸置疑，左右这一肯定的是“被”文对建筑伦理问题的诉求，而“被”文本身也在对当代中国建筑的伦理学呼吁中结束。就此而言，“被”文的积极意义不容忽视。但是，“被”文希望重建的“不作恶”“不帮凶”的所谓建筑伦理学“底限”果真应该是当代中国建筑努力的方向吗？

鉴于本文的基本主题是“中国建筑传媒奖”，笔者仍然希望围绕主题展开对上述这个问题的讨论。本届终评会结束后，终评会成员与提名和初评委员会成员以及主办媒体的代表举行了闭门会议，对“中国建筑传媒奖”举办以来的问题进行了广泛的讨论，其中论点最为集中的是如何避免“中国建筑传媒奖”成为服务社会“弱势群体”的低造价建筑、慈善基金建筑、公益建筑的专有奖项，而把大多数中国职业建筑师的工作以及决定中国城市和社会面貌的商业性和大型公共建筑项目排除在外。从已经举办的两届来看，这样的可能性确实存在。

在第一届“中国建筑传媒奖”的颁奖仪式上，“杰出成就奖”的获得者冯纪忠先生曾经就“公民建筑”说了自己的看法：“所有的建筑都是公民建筑。”笔者认为，这是一个朴素的观点，也是一个需要对其寓意进一步定义的观点。从理论上讲，社会的每一个成员都是该社会的公民，但中国的社会现实告诉我们，并非每个社会成员都在权利和义务两个方面已经是够格的公民，因此需要建设公民社会。同样，并非所有建筑都是“公民建筑”。在“中国建筑传媒奖”的语境之中，它需要符合《宣言》提出的基本宗旨并有所成就，而“中国建筑传媒奖”应该授予的，如果再次引述《宣言》的话，则是“那些关系民生，如居住、社区、环境、公共空间（注意：该范畴是上述闭门会后新增加的——笔者按）等问题，在设计中体现公共利益、倾听人文关怀、并积极为现时代状况探索高质量文化表现的建筑作品”。必须承认，相对于这些基本宗旨而言，低造价建筑、慈善基金建筑、公益建筑具有某种“与生俱来”的优势。但是，如果把“中国建筑传媒奖”的获奖作品局限于这几类建筑，便有可能使“走向公民建筑”之路越走越窄。

然而，不是所有建筑都是“公民建筑”并不意味着只有少数建筑才能成为“公民建筑”。换言之，如果把冯纪忠先生的“所有建筑都是公民建筑”在字面上略加修改为“所有建筑都有可能成为公民建筑”，那么笔者是愿意接受这样的观点的：商业建筑同样可以在“走向公民建筑”之路上有所作为，而最能体现商业建筑之“公民建筑”特性的无疑在于它的公共空间。笔者认为，这就是“中国建筑传媒奖”在其《宣言》中增加“公共空间”这一范畴的意义所在。

但是，要使所有建筑都可能成为“公民建筑”，要使商业建筑也在“走向公民建筑”之路上有所作为，仅仅遵循“被”文所谓“不作恶”“不帮

上：天津中心生态城建设公寓　　左下：新疆喀什老城区阿霍街坊改造　　右下：“土楼公社”

凶”的伦理“底限”就远远不够。在笔者看来，这样的伦理“底限”与其说是我们时代的“LESS IS MORE”，不如说是我们时代的“LESS IS SORE”更为恰当。就“被”文中出现的两个“境外建筑师”的作品鸟巢和国家大剧院而言，笔者赞同它的观点：它们在城市意义上的公共空间是失败的（尽管它们都是公共建筑），因而不能成为“中国建筑传媒奖”应该肯定的“公民建筑”（事实上也没有）。相比之下，如果我们还将自己局限在“境外建筑师”在中国的作品范围内讨论问题，那么在笔者看来，霍尔设计的深圳万科中心就是一个相对较好、甚至好很多的建筑作品（参见本书第6篇《都会田园中的建筑悬浮——评斯蒂文·霍尔的深圳万科中心》），尽管它是一个商业性质的建筑。笔者希望并相信，会有更多类似深圳万科中心这样具有商业性质的、但在环境、公共空间、人文关怀、以及现时代状况的高质量文化表现方面有探索、有成就的建筑作品能够成为“中国建筑传媒奖”之“最佳建筑奖”的有力竞争者。

笔者还认为，在“走向公民建筑”的道路上，建筑师的力量是远远不够的，它需要公众、甲方、政府等方面的共同努力。本届终评会的大多数评委都曾有这样的感觉：本届“居住建筑特别奖”不应仅仅是授予天津中心生态城建设公寓和新疆喀什老城区阿霍街坊保护改造这两个项目的建筑师，而且应该是这两个项目的组织者和其他参与者。换言之，由于这两个项目都是政府为主导的项目，本届“居住建筑特别奖”同时也应该是一个“政府鼓励奖”，就像“土楼公社”获得的第一届“居住建筑特别奖”同时应该是一个“甲方鼓励奖”一样，“中国建筑传媒奖”应该成为鼓励和促进各级政府和甲方在“走向公民建筑”上有所作为的一种力量。

“被”文说对了，在最后这一方面，大众传媒“可以凭借强大的传播平台和娴熟的操作技巧，在短时间内整合庞大的社会资源并攫取社会集体注意力，令类似‘公民建筑’这样的概念广植人心。这种能力，是整个中国建筑界和所有建筑专业媒体加在一起也望尘莫及的”。只是，在笔者看来，已有的两届“中国建筑传媒奖”还远没有做到这一点。

图"[illegible]"[illegible]，[illegible]。这种[illegible]"[illegible]"[illegible]"LESS IS MORE"[illegible]"LESS IS SORE"更为恰当。[illegible]"[illegible]"[illegible]国家大剧院[illegible]是大败的[illegible]"中国[illegible]"[illegible]之中，如果我们[illegible]自己[illegible]"[illegible]"在中国的[illegible]看来，[illegible]

[illegible]有更多[illegible]

[illegible]

"[illegible]"[illegible]

整个中国[illegible]还没有做到这一点。

13

《历史语境下关于南京博物院大殿设计的再思》点评 [1]

1　本文是为 2014“清润杯”大学生论文竞赛硕博组三等奖获奖论文《历史语境下关于南京博物院大殿设计的再思》写的导师点评，最初发表于《建筑的历史语境与绿色未来——2014—2015“清润杯”大学生论文竞赛获奖论文及点评》，《中国建筑教育》编辑部编，北京：中国建筑工业出版社，2016。

20世纪初，随着中国社会现代化进程的开始，中国现代建筑学从无到有，逐步建立起来。但是与世界上许多国家和文明的现代化不同，中国社会的现代化进程是以持续断裂为特点的，战争、国家政权的暴力更迭、政治运动，意识形态革命等，这些都导致了中国现代化进程的断裂特征。

《建筑的历史语境与绿色未来：2014—2015“清润杯”大学生论文竞赛获奖论文点评（2016）》

这一特征在中国现代建筑学的发展过程中也尽显无遗。作为中国现代建筑学的创始人和积极参与者之一，梁思成先生孜孜以求的现代中国建筑学是一个从传统向现代有机过渡的建筑学，是一个积极吸收世界先进文化同时又深深扎根于本土优秀传统的理想。1935—1947年间建成的南京国立中央博物院（现南京博物院）大殿是梁思成先生上述思想的集中体现。然而，梁思成先生的理想不仅需要对世界“先进”文化和本土“优秀”传统的甄别，而且需要一种相对持续的社会和文化环境才有可能得以实现。遗憾的是，梁思成先生的现代中国建筑学探索随着日本侵华战争以及之后中国全面内战的爆发被迫中断（尽管国立中央博物院大殿本身最终还是建成了）。随着新中国的成立而重新焕发的希望又在此后一轮又一轮的政治和意识形态运动中被无情击碎。梁思成先生最终也在无所适从的彷徨之中结束了他的现代中国建筑学之梦。

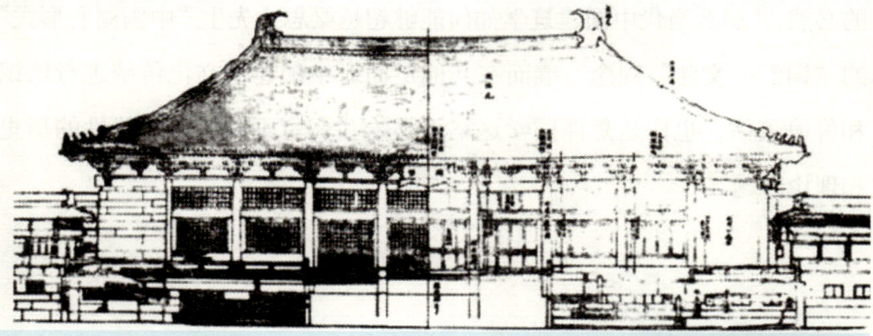

南京国立中央博物院
上：设计方案立面图　　下：现状

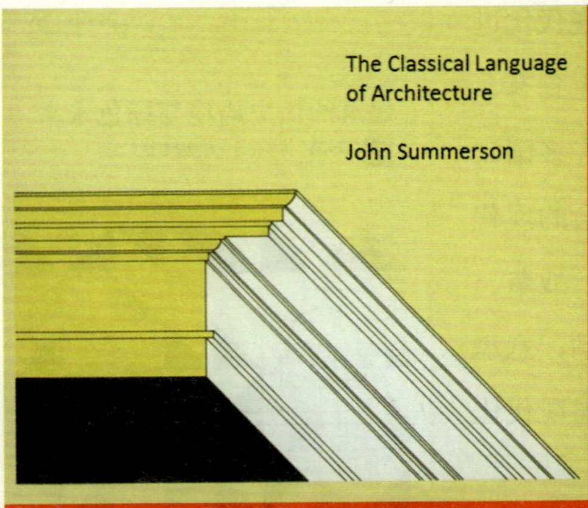

约翰·萨默森　左：《建筑的古典语言》英文版　右：《建筑的古典语言》中文版

尽管如此，在21世纪的今天，重温以梁思成先生为代表的现代中国建筑学思想仍然具有重要的理论和现实意义。焦洋同学的《历史语境下关于南京博物院大殿设计的再思》对梁思成先生建筑思想在历史和理论两个层面展开研究。在历史层面，焦洋同学有效展现了梁思成建筑思想的历史语境；而在理论层面则着重分析了梁思成先生在"中国固有形式"框架下于"美术"和"结构理性主义"之间的思想徘徊。特别是全文最后将上述研究与梁思成先生关于"词汇—文法"的思想联系在一起，应该说这在一定程度上触及了问题的核心，因为即使在现代化进程被认为最内发于自身文化传统的西方，现代建筑也是一场史无前例的变革。就此而言，曾经以《建筑的古典语言》（*The Classical Language of Architecture*）一书著称的英国学者约翰·萨默森（John Summerson）在1950年代竟也发出"逝去的语言终将逝去"（the missing language will remain missing）的感叹并坦言当代建筑学在"古典语言"以外寻求发展的必然，[1] 那么当代中国建筑学如何能够超越梁思成先生"中国固有形式"的"词汇—文法"观念，继而在其他层面对中国建筑文化传统进行认识和价值重估，也许是焦洋同学这篇论文给我们留下的最具挑战性的历史和理论问题。

1 John Summerson: "The Case for a Theory of Modern Architecture" in *Architecture Culture 1943-1968* (Edited by Joan Ockman), New York, Columbia University and Rizzoli International Publications. Inc, 1993, pp.227-236(235).

14

关于大舍建筑的几点见解[1]

1 本文原为 2017 年 1 月 21 日在上海半层书店举办的“即物即境 | 建筑论坛暨《UED 大舍专辑》新刊发布会”上的发言，收录本文集时有修改。

在今年年初举行的《建筑学报》新一届编委会的发言中，有数位编委使用了“体制外”的表述，缘由是《建筑学报》新一届编委中增加了诸如大舍的柳亦春和都市实践的王辉这样的“体制外”编委。“体制内”与“体制外”之说不是首次出现，但是在《建筑学报》这样比较权威的会议上被谈及，据我所知还是第一次。当然，我们无需对这一说法做过度诠释，所谓“体制内”指的是大型国有设计院（改制与非改制的）以及依附于高校体制的建筑师，而“体制外”则是从这一体制分离出来的或者从来就没有进入这一体制的个体建筑师。无论如何定义，我们有理由认为，“体制外”建筑师构成了当今中国建筑最好的同时也是最具创造性的建筑师群体，而大舍则是这类建筑师中的优秀代表之一。

UED 专辑《大舍：即物即境》

经过近 20 年的建筑实践，大舍不仅积累了令人刮目相看和具有国际声望的建筑作品，而且更为难能可贵的，它一直是当代中国建筑中最勤于思考的建筑师之一。为了给自己的设计找一些说法，当代中国建筑师纷纷提出一些像是设计理念的东西。但是大多数情况下都是些似是而非的修辞，有些甚至比较搞笑。相比之下，大舍的思考由来已久，而且是比较深层次的，无论是之前曾经有过的“离”“像鸟儿一样轻”，还是这次 UED 专辑中的“即物即境”。

我很欣赏“即物即境”的表述。“即物”是物自体，是建筑本身，而“即境”则渴望超越物自体，超越建筑本身；“即物”是“零度”（坂本意义上的零度），“即境”则意味着温度的回升；“即物”是结构，“即境”则是结构何为，是构架下的空间，与身体的关系，场所的意义；“即物”是“内在的结构”，“即境”则是“外在的风景”。

中国当代建筑学对“建构”问题的问题已经有近 20 年的时间。在这个问题上，我一直试图强调结构对“建构”的重要性，这无疑针对中国现代建筑对结构和建造不够重视的普遍现象，或者即使关注了结构，也只是把结构视为建筑造型的手段，这种对结构的认识是工具性的，失去了结构以及与之相关的建造对于建筑学的本体意义。在我看来，近年来方兴未艾的“结构建筑学”就是对这一问题的回应。但是对结构及其建造的强调容易导致一种误解，认为这是在鼓吹唯结构论和物质主义，把建筑降格为没有文化的物，以清华大学周榕老师的《三亭》为代表。在

大舍　上、下：上海西岸大舍工作室外部、内部

我看来，尽管这篇文章在一定程度上反映了中国现代建筑学中在技术维度缺失的情况下用“诗”来反对技术的思维方式和倾向，但是仍然值得一读，因为它可以引发我们在这个问题上进行更为深入的思考和讨论。

然而，这样的争论对于大舍并没有太大意义，因为大舍体现了中国当代建筑（至少是“体制外”的建筑）中为数不多的既对结构和技术有精深的理解和驾驭，又努力超越技术的建筑实践，“即物即境”就是对这一实践的诉求的最好表达。

如何超越物质？毫无疑问，建筑的最终目的不是物质，而是物质承载的别的什么，精神的、文化的、生活的、艺术的。但是在这方面，建筑学必须有自律，避免滑向精神和文化的过度。在我看来，进入 20 世纪之后，以民族性或者中国性为诉求的精神和文化过度是中国建筑的一个主要特征，时至今日仍然以这样或者那样的方式存在着，甚至与彻底的

大舍　左：嘉定新城幼儿园　　右：花草亭

消费主义和商业主义同流合污。在这样的语境中，我更为欣赏的其实是大舍建筑中的克制，而非过度。

非物质化是大舍建筑的另一特征。非物质化是超越物质的一种途径，在这方面日本建筑对大舍的影响不言而喻。大舍建筑也在努力寻求一种建筑中的“坂本式”“放松”，西岸的大舍工作室可以被视为这一努力的结果。不过在我看来，这个建筑仍然处在“外松内紧”的状态。

在我看来，大舍建筑总有一种若隐若现的手法主义（亦称矫饰主义）倾向。按照建筑史学家佩夫斯纳的经典定义，手法主义建筑的表征之一就是对结构表现力的否定以及对材料的操弄，以便让它看起来是与其自身完全不同的另一种物质。[1] 如果说早期大舍建筑的手法主义倾向主要在于对面纱式的建筑表皮与建筑实体以及建筑空间之间的冲突和张力的追求的话，那么近年来的大舍建筑则在关注结构的同时呈现出一种“结构手法主义”倾向。在后一个方面，或许没有什么比周榕老师“三亭”之一的“花草亭”更能体现佩夫斯纳意义上的手法主义，而台州当代美术馆大张旗鼓地使用没有柱子或梁支撑的“连续拱”——我

大舍　台州当代美术馆

1 佩夫斯纳的原文是：“Mannerism has no faith in mankind and no faith in matter⋯. The painter distorts figures to force them into imposed frozen patterns, and the architect designs with seemingly thin materials, which never look strong enough to associate with them the carring force of the human body, nor ever look massive enough to associate with them the weight of matter pressing down under the force of gravitation.” Nikolaus Pevsner, “The Architecture of Mannerism” in *The Mint: A Miscellany of Litterature, Art and Criticism*, Volume 1, London 1946, pp.116-138(135).

左：瑞士蒙特卡罗索的安蒂科修道院
右：迪艾斯特 “海鸥”加油站及服务用房

以为，这一做法不仅根本有别于以拱券见长的欧洲传统建筑的做法，而且也与乌拉圭建筑师埃拉蒂奥·迪艾斯特（Eladio Dieste）建筑中的拱结构悬挑在受力关系上不能相提并论，更不要说将“连续拱”作为建筑立面使用以视其获得面向入口广场的“公共性”的诉求了。所有这些似乎都在刻意暗示一种超越既有结构形式固有逻辑的企图。

我无意夸大这些做法的不当。我并也不认为手法主义不能出现在建筑中，尽管我也相信，手法主义只有高手才能使用——在这一点上大舍无疑当之无愧，而且即便在高手那里，过多使用手法主义也会导致腻味。我想强调指出的是，大舍最好的建筑作品是非手法主义的，我指的是龙美术馆（西岸馆）（下文简称“龙美”）。

在大舍的作品中，龙美是一个分水岭，或者说是一个巅峰之作。之前的作品是形式有余，而偶然性不足。之后的作品则试图超越龙美，但似乎仍处在超越的摸索甚至迷茫之中。

是什么使我将龙美视为大舍的巅峰之作？结构与空间乃至身体的关系是一个重要的因素，但绝非唯一因素。更为重要的也许是它突破了大舍建筑中那种过于优美的、过于干净的、有时也是似曾相见的形式，过多来自建筑师的控制，过多先入为主的意念。

就此而言，中国当代建筑中的种种粗鄙和不确定条件对于某种建筑师来说也许只是建筑品质粗制滥造的根源，而对于大舍这样才华横溢的建筑师来说反倒孕育着提升建筑品质的一种可能，前提是大舍对过程的专业驾驭以及在每一种情况下非同凡响的判断能力。

最后我愿意再次回到本次发言开始提到的“体制内”和“体制外”的话题。如果说随着建筑市场以及大型设计院工作室制度的日益成熟，“体制内”和“体制外”的差异势必逐步缩小，特别是当这种差异是以建筑作品的形式来看的时候，那么我相信在中国现行制度下，真正能够容忍和践行建筑批评的建筑师将首先来自“体制外”而非“体制内”。今天的大舍建筑座谈会就表明了这一点。

15

《建筑七人对谈集》序言[1]

1 最初发表于《建筑七人对谈集》，上海：同济大学出版社“光明城”，2015 年版。

有一种说法，中国是世界上最大的建筑工地，是建筑量的巨人，却是建筑思想的侏儒。也许，这样的论断并没有经过什么严格的论证，有信口雌黄之嫌。不过，与其纠缠这一论断的对与错，愿意对当代中国建筑进行批判性反思的人们也许会更加关注当代中国建筑数量有余而思想不足的原因，比如中国社会长期以来对建筑作为社会产品和控制工具的重视、建筑学传统的缺失、实用主义文化观，以及当代压倒一切的商品经济浪潮，等等。在这里，建筑可以是生产的手段，也是象征的符号，但绝不是思想的对象。

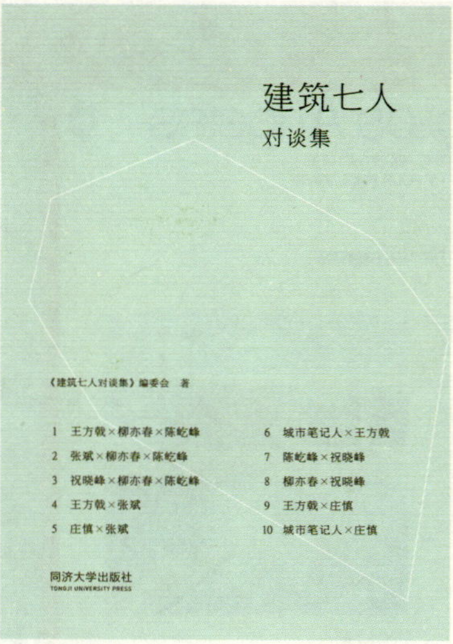

《建筑七人对谈集》

自意大利文艺复兴以来，欧洲建筑学的传统是理论先行。这种传统被 20 世纪现代主义发扬光大，以至于给人这样一种印象：20 世纪建筑学发展就是理论的胜利，尤其是宣言的胜利。未来主义、表现主义、新艺术运动、风格派、包豪斯、构成派、粗野主义、理性主义、光明城市、阿基格拉姆、后现代主义……这些思潮无一不是宣言式的思想和理论在先，然后才有建筑，至少理论与建筑同时出现。即便是原本为解决实际问题而提出的设想，如勒·柯布西耶的多米诺体系，也能摇身一变，俨然成为一种思想的偶像，透射着建筑学的理论光芒。

然而，理论先行并非建筑发展的唯一途径。在美国，资本主义的商业需求与山姆大叔的实用主义一拍即合，使 20 世纪美国建筑和城市发展沿着一条与欧洲现代主义截然相反的道路前行。没有理论，没有宣言，也较少有闪光的思想，取而代之的是物质、摩天大楼、郊区别墅、单调的城市网格、技术成就、装饰象征主义、低俗的商业消费设施，就连与多米诺住宅看似异曲同工、颇具理论潜力的芝加哥框架最终也止步于它的实用性而未能给建筑学更多贡献。

直到 1970 年代，来自欧洲的库哈斯重新发现了美国，更准确地说是重新发现了曼哈顿。“20 世纪末，已经对宣言嗤之以鼻的时代，如何重写一份城市形式的宣言？宣言的致命缺点是缺少佐证。曼哈顿的问题恰恰相反，满坑满谷的佐证，却没有宣言。因此，本书是一本为曼哈顿制作的回溯式宣言。”——库哈斯在《癫狂的纽约——一部为曼哈顿而写的回溯式宣言》(*Delirius New York: A Retroactive Manifesto for Manhattan*) 中如是说。[1] 与此同

1 Rem Koolhaas, *Delirious New York: A Retroactive Manifesto for Manhattan* (New York, The Monacelli Press, 1994), p.9. 也见雷姆·库哈斯：《癫狂的纽约：给曼哈顿补写的宣言》（唐克扬译），北京：生活·读书·新知三联书店，2015 年版，第 11 页。

左：班纳姆《洛杉矶——四种生态的建筑学》英文版　　右：文丘里、布朗、艾泽努尔《向拉斯维加斯学习》中文版

时，也是来自欧洲的班纳姆在洛杉矶看到了他的“四种生态的建筑学”（*Los Angeles: The Architecture of Four Ecologies*）。不久，美国本土的文丘里提出“向拉斯维加斯学习”。在库哈斯看来，作为“最后”一部建筑宣言（the “last” manifesto on architecture——指《建筑的复杂性与矛盾性》）的作者，文丘里的《向拉斯维加斯学习》（*Learning from Las Vegas*）开启了一种关于城市的回溯式宣言的思想模式，[1] 其经典表述方式是“向……学习”。在这里，理论落后于实践，但恰恰是这种落后为建筑学的迂回和突围提供了契机。

某种意义上，当代中国建筑发展的道路是美国式的，而非欧洲式的。不同的是，它仍然处在其“初级阶段”，很少有真正属于自己的技术成就，更缺少像SOM那样能够全面驾驭技术问题的建筑设计事务所。不同之处还在于，尽管库哈斯已经通过“哈佛城市研究”制造出《突变》（*Mutations*）和《大跃进》（*Great Leap Forword*）两部关于当代中国建筑和城市的“回溯式宣言”，但是这回他的曼哈顿招数却没有取得同样具有说服力的理论成就。就此而言，当代中国建筑学的理论突破仍然在“等待戈多”。这个“戈多”或者“戈多们”可能是欧洲人，也可能是美国人或者别的什么人——当然，我更愿意是中国人。

尽管如此，当代中国建筑发展有一点还是与美国的情况相似，那就是，在近乎荒芜的建筑文化之中仍然有一些建筑师拒绝随波逐流，努力追求建筑的更高境界。在美国，赖特、康、艾森曼、霍尔已经为世人所熟知；在中国，数量上更多、同时也更加年轻有为的新锐建筑师们（当然也包

1 AMOMA / Rem Koolhaas / &&& Simon Brown Jon Link, “Interview with Denise Scott Brown & Robert Venturi: Relearning from Las Vegas,” *Content* (Köln: Taschen, 2004), pp.150-157(150).

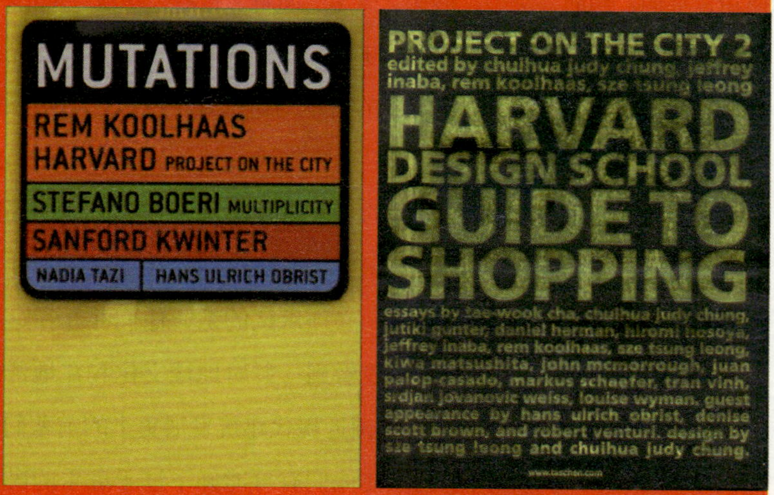

哈佛城市研究项目　左：《突变》英文版　右：《购物指南》英文版

括这本对谈集中的建筑师们）正在崭露头角，开始或者正在被世人所了解和熟知。当然，即便在这一层面上，两者之间的差异仍然显而易见：赖特、康、艾森曼、霍尔都是对建筑学有理论贡献的建筑师，而活跃在当代中国建筑前沿的新锐建筑师们却难以与他们在这一点上相媲美。

也许，正如张斌在本书的对谈中所说的，"我们不是一个精神性的民族"——因而也不是一个理论性的民族。然而事实是，在过去的几十年中，中国当代建筑学其实并不缺乏构建理论的尝试，从形形色色的"建筑哲学""建筑哲理""建筑美学"再到"广义建筑学"，建筑学的理论诉求简直可以用"辉煌"来形容。可惜的是，如此辉煌的理论诉求常常成为建筑学"不可承受之重"，令人望而生畏，直至敬而远之。

在这样的意义上，我更愿意将上述中国建筑师们的理论"退却"视为一种有意识的选择。他们不相信"宏大叙事"式的理论，更愿意在具体的建筑设计中发现自己的理论兴趣点，即使只是星星点点、不成系统。他们不相信理论是指导实践的工具，更愿意将理论与实践视为"相互平行、相互刺激、相互影响"的思想形式（王澍语）。他们不相信理论必须长篇大论，对谈、微博、杂文同样可以具有理论意义，只要它们实践的是对建筑的思考。也许，这就是眼下这本《建筑七人对谈集》产生的初衷吧。

因此，如同文丘里的《建筑的复杂性与矛盾性》（*Complexity and Contradiction in Architecture*）一样，这是一部需要紧扣建筑作品本身才能真正理解的文集。对于这些作品而言，理论概念几句话也许就能说完，只对理论概念感兴趣的人们根本无需阅读本书。另一方面，与文丘里的《建筑的复杂性与矛盾性》不同，这些建筑作品本身并不构成本书思想的全部内容。同样重要的是设计过程——如何开始设计、如何应对设计过程中的问题、如何把设计本身变成一种反思。在这里，场地条件、空间、使用、平面、剖面、类型、形式、概念、图解、材料、建造、记忆、传统、园林、身体、光、感知、尺度、与业主的关系等被一次次反复提及。正是在这一次次反复提及中，建筑学的基本问题得到呈现，成为思想的对象。

通过这些对谈，我看到一群对建筑充满热情的建筑师——当然，他们也足够清醒，能够意识到自己的局限以及建筑师作为职业的局限。通过这些对谈，我看到一群已经意识到需要在建筑的自身追求和社会责任之间恰当平衡的建筑师——当然，必须承认的是，他们并非在这一点上总是做得很好。通过这些对谈，我看到一群对地域文化和在地传统怀有浓厚兴趣的建筑师——当然，与中国近现代建筑发展史上的前辈们相比，他们今天面临的处境和问题已经使他们有可能卸下过于沉重的民族主义包袱，也更加有可能远离过于强烈的意义以及沦为符号的建筑诉求。通过这些对谈，我也看到了一群思想上更为开放、更愿意接受世界文化的建筑师——当然，他们对一切也保持着思考的距离，而不是把它们作为最新的“思潮”和“动向”来追逐。

说到“一群”，我完全无意将他们视为一个同质的整体。曾几何时，中国建筑的动向总是趋同，一有什么新的兴趣点，大家趋之若鹜。但是，透过这本对谈集，细心的读者会发现，对谈者们已经能够保持自己相对独立的思考，同时又期望在对谈中建立基本的专业共识。

更值得一提的是，“城市笔记人”刘东洋的加入不仅展现了一位“网络学者”思考建筑的另一面，而且也再一次说明，处于“体制”之外的“自媒体”能够成为推动当代中国建筑学发展的可贵力量。

最后，我愿意回到本文之前谈到的“回溯式宣言”的问题。就在中国当代建筑已经积蓄了如此多的量却仍然在等待理论突破的“戈多”之

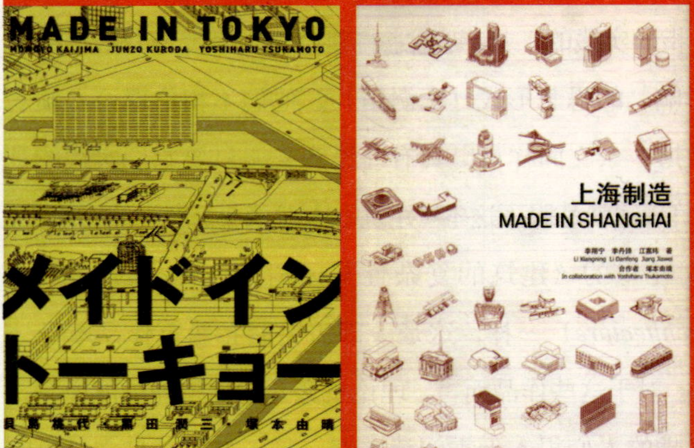

左：《东京制造》日、英文双语版　　右：《上海制造》中、英文双语版

时，塚本由晴与他人合著的《东京制造》引起了国人的注意——一部完全可以与《疯狂的纽约》或者《向拉斯维加斯学习》相提并论的“回溯式宣言”（理论）。于是，如同曾经因为“向拉斯维加斯学习”而有过“向上海学习”一样，我们也有了“上海制造”（由同济大学建筑城规学院学生在塚本由晴和李翔宁两位老师共同指导下制作的城市调研展览）。[1] 然而，我们可以有“上海制造”，甚至“广州制造”“北京制造”，但是我们似乎仍然未能将它们转变为令人信服的建筑学突破点。实质性的区别在哪里？我以为，除了换一种眼光看问题的“思想智慧”（intelligence）之外，我们缺少真正能够支撑这种思想智慧的专业水准和设计能力，而两者的兼而有之既是塚本也是库哈斯的特点，如果他们的某些建筑作品确实让我们折服的话。换言之，如果没有足够的专业技能做支撑，文丘里开启的“向低俗学习”的思想模式并不能给建筑学带来真正有价值的帮助。

柯林 · 罗《诚如我曾经所言》第二卷英文版

正是后面这一点使我们重新回到建筑学的基本问题，回到对设计能力的关注。柯林 · 罗早就指出，20 世纪“宣言文化”的谬误就在于，它假设“无需‘才华’(talent)和‘技艺’(craftsmanship)的过滤，‘理念’(ideas)会自动成为‘诗’（poetics）”。用罗在一篇载于《诚如我曾经所言》的论文中的话来说，“如果‘理念’通常折射出文化原创的想象（the fantasies of cultural primitives），那么……‘才华’就是对知识和教养的生动运用（a vivacious employment of knowledge and education）。”[2]

我以为，对于当代中国建筑学的发展而言，“理念”和“才华”一个都不能少。在这两方面，任何投机取巧和心浮气躁都不能为中国建筑的强大真正带来帮助。我衷心希望，这本对谈集不仅能够为中国建筑学的理论突破积蓄力量，而且能够为读者加深对“才华”的理解带来启迪。

1 李翔宁、李丹锋、江嘉玮：《上海制造》（合作者：塚本由晴），上海：同济大学出版社“光明城”，2014 年版。

2 Colin Rowe,“Ideas,Talent, Poetics: A Problem of Manifesto,” *As I Was Saying: Reflections and Miscellaneous Essays*, Volume Two, ed. Alexander Caragonne (Cambridge, Massachusetts and London, 1996), p.280 and p.308.

致谢

本文集以及其他几册“王骏阳建筑学论文集”收录的是本人过去二十年来陆续写成的各类文章。除了极少几篇之外，绝大多数或是应期刊杂志或出版社组稿写的专题论文，或是学术会议（包括小型座谈会）特邀发言，以及为他人书籍或文章写的序言、点评，等等，将以《理论 · 历史 · 批评（一）》《阅读柯林 · 罗的“拉图雷特”》《理论 · 历史 · 批评（二）》《建构的观点》的次序陆续出版。

在此，我要衷心感谢建筑界的同仁给我的机会，没有这些机会以及随之而来的思想碰撞和讨论，本文集以及其他文集中的大多数文章绝无可能写成。

我还要特别感谢同济大学出版社光明城编辑部的秦蕾女士对文集出版立项提供的帮助，以及对本人一再推迟提交书稿的宽容和忍耐。

同济大学建筑与城市规划学院积极鼓励和促成本书的出版，校方出版基金提供了力所能及的资助，在此一并致谢。

王骏阳

二〇一七年六月五日

时，塚本由晴与他人合著的《东京制造》引起了国人的注意——一部完全可以与《疯狂的纽约》或者《向拉斯维加斯学习》相提并论的“回溯式宣言”（理论）。于是，如同曾经因为“向拉斯维加斯学习”而有过“向上海学习”一样，我们也有了“上海制造”（由同济大学建筑城规学院学生在塚本由晴和李翔宁两位老师共同指导下制作的城市调研展览）。[1] 然而，我们可以有“上海制造”，甚至“广州制造”“北京制造”，但是我们似乎仍然未能将它们转变为令人信服的建筑学突破点。实质性的区别在哪里？我以为，除了换一种眼光看问题的“思想智慧”（intelligence）之外，我们缺少真正能够支撑这种思想智慧的专业水准和设计能力，而两者的兼而有之既是塚本也是库哈斯的特点，如果他们的某些建筑作品确实让我们折服的话。换言之，如果没有足够的专业技能做支撑，文丘里开启的“向低俗学习”的思想模式并不能给建筑学带来真正有价值的帮助。

柯林·罗《诚如我曾经所言》第二卷英文版

正是后面这一点使我们重新回到建筑学的基本问题，回到对设计能力的关注。柯林·罗早就指出，20 世纪“宣言文化”的谬误就在于，它假设“无需‘才华’(talent)和‘技艺’(craftsmanship)的过滤，‘理念’(ideas)会自动成为‘诗’（poetics）”。用罗在一篇载于《诚如我曾经所言》的论文中的话来说，“如果‘理念’通常折射出文化原创的想象（the fantasies of cultural primitives），那么……‘才华’就是对知识和教养的生动运用（a vivacious employment of knowledge and education）。”[2]

我以为，对于当代中国建筑学的发展而言，“理念”和“才华”一个都不能少。在这两方面，任何投机取巧和心浮气躁都不能为中国建筑的强大真正带来帮助。我衷心希望，这本对谈集不仅能够为中国建筑学的理论突破积蓄力量，而且能够为读者加深对“才华”的理解带来启迪。

---

1 李翔宁、李丹锋、江嘉玮：《上海制造》（合作者：塚本由晴），上海：同济大学出版社“光明城”，2014 年版。

2 Colin Rowe,“Ideas,Talent, Poetics: A Problem of Manifesto,” *As I Was Saying: Reflections and Miscellaneous Essays*, Volume Two, ed. Alexander Caragonne (Cambridge, Massachusetts and London, 1996), p.280 and p.308.

致谢

本文集以及其他几册“王骏阳建筑学论文集”收录的是本人过去二十年来陆续写成的各类文章。除了极少几篇之外，绝大多数或是应期刊杂志或出版社组稿写的专题论文，或是学术会议（包括小型座谈会）特邀发言，以及为他人书籍或文章写的序言、点评，等等，将以《理论 · 历史 · 批评（一）》《阅读柯林 · 罗的“拉图雷特”》《理论 · 历史 · 批评（二）》《建构的观点》的次序陆续出版。

在此，我要衷心感谢建筑界的同仁给我的机会，没有这些机会以及随之而来的思想碰撞和讨论，本文集以及其他文集中的大多数文章绝无可能写成。

我还要特别感谢同济大学出版社光明城编辑部的秦蕾女士对文集出版立项提供的帮助，以及对本人一再推迟提交书稿的宽容和忍耐。

同济大学建筑与城市规划学院积极鼓励和促成本书的出版，校方出版基金提供了力所能及的资助，在此一并致谢。

王骏阳

二〇一七年六月五日

图书在版编目（CIP）数据

理论·历史·批判．一 / 王骏阳著．-- 上海：同济大学出版社，2017.11

（王骏阳建筑学论文集）

ISBN 978-7-5608-7462-3

Ⅰ．①理… Ⅱ．①王… Ⅲ．①建筑学－文集 Ⅳ．① TU-53

中国版本图书馆 CIP 数据核字（2017）第 261927 号

王骏阳建筑学论文集

理论·历史·批评（一）

王骏阳 著

出 版 人：华春荣

策 划：秦蕾 / 群岛工作室

责任编辑：李争

封面及装帧设计：王骏阳

排版制作：付超 吕旻

责任校对：徐春莲

版 次：2017 年 11 月第 1 版

印 次：2017 年 11 月第 1 次印刷

印 刷：联城印刷（北京）有限公司

开 本：889mm×1194mm 1/24

印 张：10

字 数：312 000

ISBN 978-7-5608-7462-3

定 价：98.00 元

出版发行：同济大学出版社

地 址：上海市杨浦区四平路 1239 号

邮政编码：200092

网 址：http://www.tongjipress.com.cn

经 销：全国各地新华书店

本书若有印刷质量问题，请向本社发行部调换。

“同济大学学术专著（自然科学类）出版基金”资助项目

光明城联系方式：info@luminocity.cn